ANCII

General Editor: Phiroze Vasunia, Professor of Greek, University College London

How can antiquity illuminate critical issues in the modern world? How does the ancient world help us address contemporary problems and issues? In what ways do modern insights and theories shed new light on the interpretation of ancient texts, monuments, artefacts and cultures? The central aim of this exciting new series is to show how antiquity is relevant to life today. The series also points towards the ways in which the modern and ancient worlds are mutually connected and interrelated. Lively, engaging, and historically informed, *Ancients and Moderns* examines key ideas and practices in context. It shows how societies and cultures have been shaped by ideas and debates that recur. With a strong appeal to students and teachers in a variety of disciplines, including classics and ancient history, each book is written for non-specialists in a clear and accessible manner.

PHILIPPA LANG was Professor of Classics at Emory University from 2004 to 2013. Her previous work on ancient science and medicine includes *Medicine and Society in Ptolemaic Egypt* (2013) and, as editor, *Reinventions: Essays on Hellenistic and Early Roman Science* (2004).

ANCIENTS AND MODERNS SERIES

ISBN: 978-1-84885-200-6 • www.ancientsandmoderns.com

THE ART OF THE BODY: ANTIQUITY AND ITS LEGACY • MICHAEL SQUIRE

DEATH: ANTIQUITY AND ITS LEGACY • MARIO ERASMO

GENDER: ANTIQUITY AND ITS LEGACY • BROOKE HOLMES

LUCK, FATE AND FORTUNE: ANTIQUITY AND ITS LEGACY • ESTHER EIDINOW

POLITICS: ANTIQUITY AND ITS LEGACY • KOSTAS VLASSOPOULOS

RACE: ANTIQUITY AND ITS LEGACY • DENISE EILEEN McCOSKEY

RELIGION: ANTIQUITY AND ITS LEGACY • JÖRG RÜPKE

SCIENCE: ANTIQUITY AND ITS LEGACY • PHILIPPA LANG

SEX: ANTIQUITY AND ITS LEGACY • DANIEL ORRELLS

SLAVERY: ANTIQUITY AND ITS LEGACY • PAGE DUBOIS

WAR: ANTIQUITY AND ITS LEGACY • ALFRED S. BRADFORD

'The study of ancient Greco-Roman science has undergone considerable changes in recent years, and our understandings of many aspects of modern science even more so. In this engaging new analysis Philippa Lang pulls off the remarkable feat of using the moderns to offer a fresh evaluation of the ancients' work. Thus she cites modern disputes in cosmology the better to appreciate ancient speculations about the origins of things while never underestimating the different methods used, and again recent developments in evolutionary theory to throw light on ancient debates on the changes to which animal kinds are subject. Ancient science, in her confident hands, is not just of historical interest, but brought to life as a record of the acute endeavours of inquiring minds to make sense of their environment.'

Sir Geoffrey Lloyd, FBA, Professor Emeritus of Ancient Philosophy and Science, University of Cambridge

'In her new book Philippa Lang discusses the differences and similarities between a variety of ancient and a variety of modern scientific explanations for the cosmos, species, methods, motion, and health. She offers a careful and engaging argument that brings ancient thinkers face to face with moderns such as Dalton, Darwin, Einstein, and Galileo, revealing where and why ancient explanations were found wanting, and how later formulations of particular problems and questions offered new solutions and explanations for the same phenomena. She is thought-provoking on ancient scientific method and theory; on the nature of experiment and observation, prediction and results, expectations and interpretations; and on the application of ancient scientific research to real world problems, making use throughout of well chosen and clearly explained examples.'

Tracey Rihll, Reader in History and Classics, Swansea University, author of *Greek Science* and co-editor of *Science and Mathematics in Ancient Greek Culture*

To all the Emory University students who took my classes in ancient science and philosophy in the years 2002–13, with thanks.

ANCIENTS AND MODERNS

SCIENCE
ANTIQVITY AND ITS LEGACY

PHILIPPA LANG

BLOOMSBURY ACADEMIC
LONDON • NEW YORK • OXFORD • NEW DELHI • SYDNEY

BLOOMSBURY ACADEMIC
Bloomsbury Publishing Plc
50 Bedford Square, London, WC1B 3DP, UK
1385 Broadway, New York, NY 10018, USA

BLOOMSBURY, BLOOMSBURY ACADEMIC and the Diana logo are
trademarks of Bloomsbury Publishing Plc

First published in Great Britain 2015
Paperback edition published 2019

Copyright © Philippa Lang, 2019

Philippa Lang has asserted her right under the Copyright,
Designs and Patents Act, 1988, to be identified as Author of this work.

For legal purposes the Acknowledgements on p. xi constitute
an extension of this copyright page.

Cover design: Ian Ross ianrossdesigner.com
Cover image © Archimedes in his bath. Hand-coloured woodcut, 1547
(Photo by Ann Ronan Pictures/Print Collector/Getty Images)

All rights reserved. No part of this publication may be reproduced or
transmitted in any form or by any means, electronic or mechanical,
including photocopying, recording, or any information storage or
retrieval system, without prior permission in writing from the publishers.

Bloomsbury Publishing Plc does not have any control over, or responsibility for,
any third-party websites referred to or in this book. All internet addresses given
in this book were correct at the time of going to press. The author and publisher
regret any inconvenience caused if addresses have changed or sites have
ceased to exist, but can accept no responsibility for any such changes.

A catalogue record for this book is available from the British Library.

A catalog record for this book is available from the Library of Congress.

ISBN: HB: 978-1-7807-6171-8
PB: 978-1-3501-2151-5
ePDF: 978-0-8577-2612-4
eBook: 978-0-8577-3955-1

Series: Ancients and Moderns

Typeset by Newgen KnowledgeWorks Pvt. Ltd., Chennai, India

To find out more about our authors and books visit
www.bloomsbury.com and sign up for our newsletters.

CONTENTS

ACKNOWLEDGEMENTS — xi
NOTE ON DATES — xi
FOREWORD (*by Phiroze Vasunia*) — xiii

CHAPTER I: STRANGE WORLDS

The Pre-Socratics — 1
The Pigeons and the Physicists — 5
In the Beginning — 7
Nothing Can Come Out of Nothing — 9
The Uses of Infinity — 12
The Mechanics of Eternity — 16
Problems with Time — 21
Growing Seeds and Bouncing Universes — 24

CHAPTER II: STRANGE CREATURES

The Origins of Life — 28
Implications of Fossils — 33
Origins of Species — 36
The Origins of Successful Theories — 43
Plato and Primates — 50

CHAPTER III: NATURAL LAWS AND HOW TO DISCOVER THEM

The Category 'Nature'	53
Aristotle and the Modern Scientific Method	57
Objects in Space	64
Objects in Motion (I)	66
Objects in Motion (II)	73
Motion into Empty Spaces	76
Paradigm Shifts	84

CHAPTER IV: ILLNESS AND DISEASE

Medical Cultures	88
Naturalistic Medicine: An Influential Meme	91
When to Use an Amulet	96
QED?	101
The Contagion Superstition	109
Theories about Women	117

CHAPTER V: CONTROLLING THE WORLD

Rhomboids Are Forever	128
First Find Your Mathematician	129
The Rise of the Astronomers	133
Mapping the World	140
Mapping the Cosmos	145
Empire by Numbers	148
The Rise of the Machine-Makers	152
Cubes, Catapults, Computers	155
Formulas to Live by	163
Weaponise Your Way to Inner Peace	167

CONTENTS

CHAPTER VI: THEN AND NOW
   On Not Being Human   175
   Tomorrow's Pseudo-Science   181
   Hippocrates in the Twenty-First Century   188

SOME SUGGESTIONS FOR FURTHER READING   203

NOTES   211

INDEX   224

ACKNOWLEDGEMENTS

I would like to thank my editor, Alex Wright, for his persistence in the face of adversity; Geoffrey Lloyd and my previous editor Caroline McNaught for their encouragement and helpful advice. All errors and infelicities in fact or interpretation remain my own.

NOTE ON DATES

In this book dates, such as those of a person's lifetime, have usually been specified as either BC or AD. When neither are given, the date is always AD.

FOREWORD

Ancients and Moderns comes to fruition at a propitious moment: 'reception studies' is flourishing, and the scholarship that has arisen around it is lively, rigorous, and historically informed; it makes us rethink our own understanding of the relationship between past and present. *Ancients and Moderns* aims to communicate to students and general readers the depth, energy, and excitement of the best work in the field. It seeks to engage, provoke, and stimulate, and to show how, for large parts of the world, Greco-Roman antiquity continues to be relevant to debates in culture, politics, and society.

The series does not merely accept notions such as 'reception' or 'tradition' without question; rather, it treats these concepts as contested categories and calls into question the illusion of an unmediated approach to the ancient world. We have encouraged our authors to take intellectual risks in the development of their ideas. By challenging the assumption of a direct line of continuity between antiquity and modernity, these books explore how discussions in such areas as gender, politics, race, sex, and slavery occur within particular contexts and histories; they demonstrate that no culture is monolithic, that claims to ownership of the past are never pure, and that East and West are often connected together in ways that continue to surprise and disturb many. Thus, *Ancients and Moderns* is intended to stir up debates about and within reception studies and to complicate some of the standard narratives about the 'legacy' of Greece and Rome.

All the books in *Ancients and Moderns* illustrate that *how* we think about the past bears a necessary relation to *who* we are in the present. At the same time, the series also seeks to persuade scholars of antiquity that their own pursuit is inextricably connected to what many generations have thought, said, and done about the ancient world.

Phiroze Vasunia

CHAPTER I

STRANGE WORLDS

The Pre-Socratics

The city of Miletus lies on the eastern coast of what is now Turkey, in the geographical region traditionally known to the West as Asia Minor. In the sixth century BC, however, Miletus was a flourishing Greek colony in Lydia. It had extensive trading networks and channels of communication with the Mediterranean world of Greek city-states on the mainland of Europe and islands of the Aegean, as well as with other Greek foundations along the eastern Asia Minor coast, in the Black Sea region to the north, and as far south as Naucratis in Egypt. These networks were further embedded within the societies and cultures of the region, including the countries and empires of Egypt, Lydia, Persia, Babylonia and India.

Miletus was thus ideally situated to be a nexus of cultural interaction and interchange. Perhaps this is why three of its citizens are still famous as the earliest of the 'Pre-Socratic' thinkers of Greco-Roman antiquity.[1] This collection of diverse individuals, beginning with the Milesians Thales (lived c.624–546 BC), Anaximenes (active around 545 BC) and Anaximander (c.610–546 BC), over the course of one or two generations constituted an intellectual revolution within the Greek world, a century or so before the inquiries of Socrates of Athens (469–399 BC) marked a subsequent divergence and evolution in what came to be called philosophy.

'Pre-Socratics' is, obviously, a later term of convenience, not a contemporary category. We are also handicapped in reconstructing or understanding their ideas by the fact that Pre-Socratic writings exist only in quotations or paraphrases ('fragments') of them by other, later authors, a group that has been itself winnowed by the selective and destructive passage of time. Thales, traditionally the first Pre-Socratic and therefore now occasionally identified as the first 'scientist' or 'philosopher' in Greco-Roman history, may not have produced any written argument at all. Aristotle, discussing Thales' ideas in the fourth century BC, seems to be somewhat uncertain of what these might actually have been. This suggests a reliance on oral tradition. Moreover Aristotle interprets not only Thales but all Pre-Socratics in terms of his own philosophical terminology, reflecting the development of concepts and arguments in the years since. The result is that we gain our own knowledge of their arguments through this anachronistic lens. And Aristotle is one of the earliest extant interpreters of the Pre-Socratics. Many of our sources for Pre-Socratic thought are much later than him and some were themselves relying on intermediate texts and reports.

The Pre-Socratics had sharply different ideas from each other, but they also had some things in common. They each articulated, in their own way, a radical departure from older kinds of ideas about the origins and nature of the world and its inhabitants, from gods to humans. They reformulated the very idea of what gods could be and how such beings might relate (or not) to events, processes and objects in the world around them. They made explicit and problematised questions of what constituted knowledge, persuasive argument or proof. They claimed confidently to offer a privileged and superior understanding of important elements – moral, physical, epistemological, civic – of the world. Crucially, they often offered reasons and arguments for believing in each of their particular visions. In so doing they reframed both religious and non-religious discourse to a significant degree.

Argument of this type became a very influential trend in Greek and subsequently Greco-Roman society. This was particularly so among the more elite and educated members of that society – those with the leisure to debate, the ability to read and intellectual fashions to keep up with – but had implications for culture much more widely. Philosophy, as it became known, developed as a broad genre of argument, investigation and analysis.

It influenced emerging literary genres like ethnographic history and included an evolving specialisation into subject matter that we would now consider 'scientific'. The impact of this kind of thought is particularly notable among professional healers as the form of medicine known as 'Hippocratic'.

What began as interlocutory speculation and criticism amongst the Pre-Socratics and their audiences became a dominant mode of Greco-Roman culture, both oral and written, and survived into the Christian and Byzantine eras. It affected Christian theology, political thought, literature and other arts and medicine. In spite of a considerable loss of texts throughout classical antiquity and the subsequent Dark Ages, some treatises were preserved in the Arabic intellectual tradition and rediscovered by the West in medieval times. These included the astronomer Ptolemy's *Almagest* and several works by the physician Galen. The intellectual upheavals in science and philosophy of the Renaissance, during the fourteenth to seventeenth centuries, were identified by their participants as descended from the natural philosophy of Greece and Rome. The names and works of Ptolemy, Galen, the mathematician Archimedes and the philosopher Aristotle became authorities and exemplars within this resurgence of themes from pre- and early-Christian literature and ideas. In general, the ancient practice of investigative and critical thought, concerning a rational world of consistent and understandable 'nature', became an ideal which drove the development of science and natural history, especially in the so-called 'Enlightenment' of the seventeenth and eighteenth centuries. Pushed as far back as specific individuals and ideas can be identified, it is a story that leads to the Pre-Socratics and to sixth-century BC Miletus.

As a reading of antiquity, such a story of scientific discovery is highly selective and decontextualised. As a story of scientific progress and rediscovery, it is doubtful: Greco-Roman inquiries into scientific subjects would rarely if ever – outside mathematics – meet modern criteria for what constitutes science. In the rest of this book, we shall examine how we understand and define such 'ancient science' through our own familiarity with its successful modern analogue. But the differences are also important in thinking about what the sciences are, or what or 'science' is, and how society and culture affect forms of investigation and argument.

Pre-Socratics and their next generation successors, the so-called sophists, tended to think big, about such topics as the origin of the cosmos or its fundamental nature; the nature of reality; the definitions of truth; whether gods existed and what they might be. Such thinkers were often at variance with traditional ritual, material or poetic representations of the gods; they also sought to define morality and prescribe political structure.

Their ideas as to all these differed, often radically, from each other. Many offered an explanation of the cosmos in terms of materials: a move taken to its extreme by the atomists, whose discrete, irreducible matter competed with the conception of an infinitely divisible substrate as put forward by Anaxagoras and later Aristotle. For Parmenides, on the other hand, reality was something else entirely. Matter, like movement, was an illusion. His view was taken up by Zeno and Melissus, and became a major influence on Plato's thought.

But on a broader scale, there is a common theme to Pre-Socratic ideas. All describe a hidden, fundamental set of explanatory principles or items, accessible primarily by thought and not directly perceptible by the senses or everyday experience of the world.[2] This reality is very different to the macroscopic experience it creates: we do not perceive the world as a set of interacting solid particles in emptiness (the atomists) or a table as something that is also bread and stone (Anaxagoras). We do not think of the world and everything in it as a mass of air in various stages of rarefaction (Anaximenes); we do not think of a woman being reducible to the number two, or of retributive justice as the number four (the Pythagoreans); we allow, unlike Parmenides, that change, motion and different things exist. In its strangeness, the Pre-Socratic universe is similar to that of modern science. Contemporary theories of particle physics and the quantum world offer a reality that is, if anything, even more at variance with our everyday experience and common-sense understanding than were ancient ideas. The differences lie in the methods of investigation and the evidentiary criteria of persuasion.

In the rest of this chapter we shall examine early Greek speculation and argument about the nature and origin of the cosmos. Some elements of this have notable points of comparison with aspects of modern cosmological theory and debate, particularly in the kind of questions and problems encountered in trying to think about the beginning of existence, but less so in how such issues are resolved.

The Pigeons and the Physicists

At the beginning of the 1960s, scientists at Princeton were looking for evidence of the cosmic radiation that theory predicted would have been produced in the 'big bang'. If the hot big bang model was correct, the universe – time, space, matter and energy – came into being 14 billion years ago, as something extremely hot and incredibly dense. There was no 20 billion years ago.

At the time the alternative cosmological model was the 'steady-state' hypothesis, according to which the universe was essentially the same – having the same properties – over time as well as space. Matter did not come into being at one moment, but more of it arrived very slowly in a process of 'continuous creation' as the universe expanded (as observations of stellar phenomena had showed it was doing), thus maintaining the overall universe at the same density. The amount of matter required was on such a small scale that it would be very difficult to observe and therefore very difficult to disprove. But since steady-state theory predicted that the universe's properties had not changed over time, the hypothesis could be tested by searching for evidence that the past and contemporary universe differed in key aspects. Any evidence of this would falsify steady-state theory.

The aim of the Princeton scientists – Dilke, Peebles, Wilkinson – was to look for such evidence in the particular form predicted by the big bang model, according to which the early universe was too hot for atoms to exist.

Protons, electrons and photons were scattered equally throughout, so that the emission and absorption of energy balanced each other in perfect equilibrium: a state of affairs known as 'black-body radiation'. At something over 379,000 years after the big bang, the universe had cooled down sufficiently to form neutrally charged atoms, which could not absorb all the thermal energy as perfectly as before. Instead, high-energy photons began to travel on their own through space, rendering the universe transparent instead of opaque. As the universe expanded over the subsequent 15 billion years these photons lost energy and red-shifted. Because this was energy generated in the big bang from the entire early universe (not from only one part), it should appear uniform across all of space. Finding such a uniform spread of energy, a relic of the very early universe, would indicate that the universe had been very different in the past from its current state, falsifying the steady-state hypothesis and providing strong support for the truth of the 'big bang'.

As it happened, this relic radiation was discovered by two other scientists looking for something quite different. Arno Penzias and Robert Wilson were using microwave-detecting equipment designed to find sources of interference with the earth's satellite communications. The signals they were looking for were very faint, so they needed to eliminate all other sources of radiation – such as their receiver's heat, which they cooled with liquid helium to just above absolute zero. But they could not get rid of an unexpected, persistent and invariant radiation at a wavelength of 7.35 cm, which was uniform in every direction of the sky over 24 hours a day.

They checked their equipment. There were pigeons living in the telescope antenna. Surely this unusual problem explained their unexpected readings? But removing the birds and their droppings had no effect upon the persistent long-wavelength radiation. It was not an instrumental problem.

Penzias and Wilson had found the heat energy left over from the big bang and now only a few degrees above absolute zero: cosmic microwave background radiation (CMB). In spite of gaps and imperfections, the big bang model is now standard cosmology.

In the Beginning

The big bang theory models the expansion of space–time from a tiny kernel of very high density and temperature. It is not precisely about the origin of the universe, although hypotheses concerning this also exist, but does describe how the universe in which current physical laws apply came into being. In the first few fractions of a second of this expansion existence was only on the quantum scale and then, as matter and energy separated, a matter of particles. Eventually, about a second after the expansion began, subatomic particles combined into nuclei. Classical physics describes the situation from about 0.1 of a second after the expansion began – 'the big bang' itself. Particle accelerators attempt to recreate the conditions just prior to that.

Whatever the initial state or singularity was, or how it came about, cosmic background radiation and other evidence confirms that the universe began, if not literally from 'nothing', then from something completely different, and that there has been a finite amount of time between that beginning point and the present. In antiquity, a similar debate emerged within philosophy between the same basic alternatives, but was never similarly resolved into general agreement.

There are other points of comparison as well. The issue of whether time existed before the cosmos was discussed at least as early as the fourth century BC, in Plato's cosmological dialogue the *Timaeus* (written perhaps c.360 BC). Several schools of thought developed the notion of a cosmos or world that came to an end but began again: an idea that is to some extent paralleled by possible outcomes to our own universe. There are less significant similarities: the imagery of a small homogenous seed or kernel that expands into the large and differentiated cosmos is present in both ancient and modern writing. The chief difference in this case is that in antiquity, the image is sometimes literal. I shall return to these similarities in more detail later in this chapter.

As to the question of whether the cosmos has always existed, and in much the same state as it is now, or whether it had a beginning, this was formulated clearly by Plato (427–347 BC).

> Concerning it [the world], then, we must start with the question which is laid down as the question one should ask at the beginning of any topic, whether it always was, having no beginning point of coming into existence, or whether it has come into being, beginning from some beginning point. [In fact] it has come into being.
>
> (*Timaeus*, 28)

The Greek noun *genesis* means, literally, 'state of having-come-into-being', and this meaning is non-committal as to whether it refers to a process of creation by an external agent or was self-caused without plan, consciousness or agency. Early Greek society contained stories about origins of gods, men and the world. There was no fixed religious doctrine on any of these points, but the famous and very influential poem *Theogony*, composed by Hesiod in about the beginning of the seventh century BC, is still extant. In the *Theogony* or 'Birth of the Gods', the original deity is Chaos, from which all the other gods and forces derive in an expanding genealogy, starting with Earth. The meaning of chaos is space or void, as in the gaps between things, but it also implies a material fluidity or flux.

The Pre-Socratics were clearly influenced by this kind of narrative. Their cosmogonies tend to begin with some primary element such as water, air or fire. This basic material undergoes various changes and differentiations in order to produce the cosmos, consisting of the world at the centre, the sun and other wandering stars (planets) and the fixed stars. Further evolution of landscape and life then occurs upon the earth. Anaximander of Miletus' version of this stuff from which everything else came was 'the unlimited', or 'that without boundaries': something that was potentially anything but, actually, no particular thing – a more precisely defined version of Hesiod's chaotic potential. In the fourth century BC the divine craftsman of Plato's *Timaeus* is said to have crafted the world's basic material elements (cubes, icosahedrons, octahedrons and tetrahedrons, all made out of certain kinds of triangles and constituting respectively earth, water, air and fire) out of the chaotic motions of what Plato calls 'the receptacle'. This last is a kind of metaphysical substrate of matter which a divine craftsman can organise into the elemental materials of the physical universe; and it too shares a conceptual similarity with Hesiod's chaos.³

The original, basic meaning of the Greek word *kosmos*, Latinised as cosmos, is 'order, what is regular'. It could and did refer to many kinds of things which had been converted from something formless into something with structure. In the latter sense it could even mean 'hairstyle'. But at some early stage in Greek thought it became the standard terminology for the entirety of the world, including the celestial realm. The Greeks thought of the world or universe as something intrinsically ordered and hence explicable. In origin stories such as the *Theogony* or the more material accounts of the first Pre-Socratics, this ordered state of being – the universe as we know it – has a beginning, but it is not a beginning out of nothing. A previous state, of indefinite or infinite length, incomprehensible to explication and reason because it entirely lacked structure, is the apparently necessary precursor to genesis. *Something must come first*

There is a similarity here with the big bang model: the explicable form of the universe, as described by classical cosmology, has an origin point. But it did not emerge from nothing. Instead it was the result of a different kind of physics, the physics of a singularity, that is of matter-energy behaving so differently in extreme conditions as not to be matter and energy as they currently exist and are described by physics. Yet modern cosmology does offer hypotheses as to what the conditions could have been in the first few moments of the 'big bang' or even at its beginning, in the form of quantum physics and perhaps 'quantum tunnelling'.⁴ Although a singularity is by definition pretty inexplicable, more can be said about what is and is not implied by the equations describing a black hole than was the case for chaos or Anaximander's 'the limitless'.

Nothing Can Come Out of Nothing

At the same time, however, the notion of the cosmos having a beginning at all was being problematised even as it was made explicit. In the earlier decades of the fifth century BC, the Pre-Socratic Parmenides (*c*.515–450 BC) articulated a detailed rejection of the possibility of creation *ex nihilo*. His formulation brought it to the front and centre of Greek philosophy, creating one of the defining tenets of thought on all kinds of issues for the next thousand years.[5] 'Nothing', said Parmenides, 'comes out of nothing.'

Parmenides presented his thought in poetry, which for Greek culture was the traditional mode of inspired truth and cultural authority; he also used metaphor as he described how a goddess was his guide to the paths of seeming and understanding and the difference between the two. His arguments are however complex and difficult enough that, even had he employed prose, we would probably have had equal difficulty in discerning his exact meaning. This is not the place for an exploration of it, but the short version is that Parmenides is arguing for (some variety of) monism, the theory that what exists is not many things (tea, life, hydrogen, the Battle of Hastings) but one thing that is undifferentiated and eternal. Change is the transformation of something (let us call it X) into something else (let us call it Y) and in a sense, then, Y has come-into-being out of nothing. Which is impossible.

It is not clear whether Parmenides was arguing that change has no reality at all (strict monism) or whether he thought that there was a different kind of existence for temporary things – the state of being hot, an individual life – the things which we perceive as the components of our universe but which we cannot understand securely because they are contingent and insecure in their identity/existence.[6]

In either case, however, there is a different kind of reality: it is single, singular, uniform and unchanging, and in existence forever without a beginning. Knowledge of this is the true understanding; knowledge of mundane pluralities is only a contingent, imperfect grasp of what seems, but is not really, to be the case. On this approximation, Parmenides was allowing for a kind of understanding not unlike what we call 'science', and it was perhaps in this spirit that the second half of his poem offers a cosmological account of its own. But he was also arguing that there is another truth which can only be approached by a different means – a kind of inspired logic – and that this is clearly the superior version.

Monist metaphysics influenced Plato heavily, but even those who disagreed with the consequences derived by Parmenides had to take account of his arguments. They had to explain how change in time and motion through space (since motion is a change of place) could be real. The concept of nothing itself was also explicitly problematised as something that could not, as it were, be anything, so that inconceivability and impossibility became linked. Again, ideas of some kind of Hesiodic chaos or Anaximander's 'the limitless' did not necessarily solve this difficulty.

In antiquity, then, the question of whether the universe had an origin became first and foremost a problem of logic within the arena of ontology. It was also a significant factor in developing or arguing for the various theories of 'physics' on offer from the Pre-Socratics and their intellectual successors. And it gave cosmologies involving some kind of genesis event a particular problem: how to explain what had happened without falling victim to Parmenides' rule.

As we have said, the invocation of a pre-cosmos kind of chaotic potential, for which normal explanations of cause, matter and reality did not apply, has a structural similarity to the pre-classical conditions of the big bang. But the questions are not focused in quite the same way. In antiquity the question of whether the cosmos had a beginning was a relatively isolated and central point of dispute, albeit one whose answers expressed something fundamental. The arguments surrounding it were also more purely logical in nature than those of modern cosmology. They had far less to do with observation and were almost entirely unrelated to mathematics, although concepts of the finite and infinite were in play. In the Pre-Socratic period there was little consensus as to what a fundamental physics might consist of or how astronomical phenomena might be explained. Explananda, explanations, techniques and data were all in much shorter supply than in the modern world, producing a relatively limited set of questions and their potential answers.

For cosmologists of the mid-twentieth century, the issue at hand was which of two competing cosmological models was the best explanation for the universe as they understood it, both in its current state and including the history that was visible through the observation of its well-travelled light and radiation. Both models, not to mention then-current alternatives such as the 'cold big bang', were constructed out of pre-existing extensive and complex observational, experimental and theoretical physics and cosmology, about which there was a very large degree of consensus. Both, for example, involved a knowledge of atomic physics, the nature of light radiation, thermodynamics; both operated in an Einsteinian universe and had to be compatible with his equations; both had to take into account the fact that the universe was expanding, as observations indirectly demonstrated.[7] The question as to whether the universe had or did not have an origin point was certainly one of the arguments involved, but it was ultimately a subsidiary issue dependent on the success of these larger cosmological models and decided by the conclusive observation of the CMB predicted by big bang theory.

The Uses of Infinity

One obvious solution to the conceptual difficulties of a universe beginning out of nothing, or out of something so different as to be inexplicable, is a universe that did not begin. Versions of a steady state model had adherents in antiquity also, among whom was Aristotle (384–22 BC). His vision was of a spatially finite cosmos, consisting of the world at its centre and the circular motions of the celestial spheres, that had always existed and always would. A cosmos infinite in both time and space was also an integral part of atomic theory as developed, first, by the Pre-Socratics Leucippus and Democritus and, subsequently, from about 300 BC onwards, by the followers of the philosopher Epicurus.

Aristotle argued that the cosmos has always existed without a beginning. His grounds were the argument that motion must be eternal: 'That there never was a time when there was not motion, and never will be a time when there will not be motion' (*Physics* 8, 252b6–8). His reasoning is that nothing that is, itself, without motion, could produce motion either in some part of itself or in something else. In Aristotelian cosmology the outermost sphere of the heavens, in which the non-planetary stars are located, moves in a circle, which unlike rectilinear motion is continuous and undifferentiated because it has no starting-point, middle or end. It is therefore eternal (*Metaphysics* 12.7, 1072a24) and the contiguous cause of all other motion in the cosmos, this being transmitted from the outermost celestial sphere through the inner regions of the cosmos and then to the terrestrial domain. (It is perhaps worth pointing out here that to any terrestrial observer the stars appear to move around a stationary earth, and this was accepted as a cosmological fact by both casual observers and philosophical theorists.)

The eternal motion of the first heavenly sphere must also have a cause, but to avoid an infinite regress, that cause cannot itself be in motion. In explaining how something not itself in motion can be the cause of motion in others, Aristotle reasoned analogically from the domain of human activity to the cosmos. If I desire a drink of water, what causes my going to the kitchen for that drink is my desire to be not thirsty or, if you like, my wish to feel better. Similarly, going to philosophy school, for Aristotle, is caused by one's rational desire to be good. He draws the parallel conclusion that the unmoved eternal actuality causes motion by being an object of desire for the heavenly beings of the outermost sphere: 'The final cause, then, produces motion as something that is loved, but all other things move by being moved' (*Metaphysics* 12.7).

The deep structure of Aristotle's cosmos is not physical laws of behaviour, but what might be called 'values', though that is not all they are. In Aristotle's cosmos, things are the way they are because they have reason to be so, that is there is value in their being that way rather than another.

This is easier to think about in terms of an intelligent, conscious designer: Plato had described such a being as a kind of craftsman and Aristotle often uses an analogy with craftsmanship and craftsmen in discussing the workings of nature. There are aspects of Plato's cosmological analysis that Aristotle is in at least some agreement with and among those is the conviction that the cosmos is well-designed, fit for purpose, rational and good. The difference is that Aristotle eliminated the designer: the fully achieved cosmos is its own reason for existence and hence the cause and reason for how it comes to be fully achieved. The cosmos, and everything in it that happens on a regular basis (leaving out the accidents of luck, such as a sapling struck by lightning and prevented from becoming an oak), is teleological: everything happens for a reason. In Aristotle, the reason for being an oak sapling is to become an adult oak and the characteristics of an oak – the shapes of its leaves, its materials, its height – are caused by, as well as causes of, its adult nature.

Aristotle assumes that some options for existence are to be preferred to others as better and superior. Nature, he thinks, inherently moves in those directions when it is possible to do so (the craftsman is only as good as his tools). Even certain literal directions are privileged: right over left, up over down, affecting such issues as the placement of the heart and human bipedalism. The circular motion of the celestial bodies as described above is the most perfect form of motion because a circle is the most homogenous shape and so motion in a circle is an unchanging kind of motion. That makes it the form of motion most similar to not being in motion at all, and it is the latter which the celestial bodies emulate to the extent permitted by their nature. They move in circles because of their natural drive to be as close to perfection as possible and the unmoved mover is situated on the outermost edge of the cosmos as an infinite being without any magnitude. In its perfection and its primacy, as the cause of the motion of others, the unmoved mover is Aristotle's definition of god. The best of the cosmos is instantiated as divinity, but its agency consists simply of being there to be emulated.

The term nature, by which we translate the Greek word *physis*, is Aristotle's shorthand for the motive arrangement of matter, space and time that enables items within the universe to be the best they can be and for the cosmos as a whole to be the best possible. Humans rank quite far up on this scale, but the ultimate expression of natural perfection is god. What we think of as the 'laws of physics' or of nature are some of the recurring patterns in this arrangement, as the means by which that which exists orders itself in the most suitable way possible.

Aristotle can appear very modern in his approach to the natural world. He rejected a designer but emphasised the role of function in biology, being an early pioneer in explaining how different animals' anatomy and physiology worked for their species' needs and capacities.[8] In showing this he utilised empirical investigation to a far greater extent than any before him (and most after him, until at least the seventeenth century) and more or less founded empirical science as an epistemologically justified pursuit. Yet we can see from the brief summary above that this apparent modernity comes from a different place than actual modern science and negotiates a different terrain of contemporary debate, problems and assumptions. The nature and role of divinity is completely relevant to cosmology and biology but not in the ways in which it appears in modern argument. Aristotle's concept of a teleological nature, a concept which had huge explanatory power, was predicated on the assumption that the cosmos and its constituents had an intrinsic direction and a natural hierarchy.

The terms of this debate had been most recently set by Plato, but Aristotle was also dealing with another element in the mix. The atomist Democritus (*c.*460–370 BC) had also argued for an eternal universe with no beginning, but his conceptual framework and the consequences drawn from his argument were very different from Aristotle's.

For the atomists the world was an accident of past physical events, with no more intrinsic value to it than any other possible outcome. This was anathema to Aristotle, whose development of a cosmos that in spite of being unplanned nevertheless necessarily exhibited a rational perfectionism, as if designed, was devised in opposition to the contingent determinism of Democritus' universe. But in the atomic conception, infinity has quite a different set of implications. Instead of demonstrating in its everlasting lack of change the perfect and unchanging eternity of higher forms of nature, the virtue of infinity to the atomists was that it allowed for all kinds of change to happen.

The Mechanics of Eternity

According to the atomists, the irreducible constituents of the cosmos are atoms, unbreakable solids (*atoma* literally means 'uncuttable') of varying sizes and shapes that are continually falling through emptiness, or void.[9] Everything else, including worlds, humans, salamanders, blood and grass consists of accidental assemblages of atoms which have collided with each other on their trajectories through the void and in some cases have stuck together instead of bouncing off separately again. The Roman poet Lucretius compared the movement of atoms to dust particles visible in a shaft of sunlight.

Atomic complexes make up the macroscopic universe, including its perceptible qualities such as temperature, colour, sweetness and so on. These are not a property of any atom on its own, but are epiphenomena resulting from the interaction of the particular atomic complex of a human or animal sensory system with those of other atomic complexes around them, making a leaf seem green to us or honey sweet. Greek philosophers often drew attention to the fact that different people perceive something differently: of two people standing next to each other, for instance, one might find the wind colder than the other. This was sometimes used as a point in favour of relativism, as it was by the fifth-century thinker Protagoras (*c*.490–20 BC): the wind is neither cold nor warm in itself, it is only cold or warm for the person who experiences it that way.[10]

But the atomist explanation is different from relativism, because the atomic structure of the item experienced as blue, cold, wet or salty is still part of what causes that perceptual experience. Because it is also an experience of the perceiver and is determined partly by that perceiver's internal atomic arrangements, Democritus' idea works as an explanation for something the Greeks guessed might be the case but did not know: the fact that different species perceive the same things differently. Bees, for instance, can perceive the ultraviolet end of the light spectrum, so that for them a plant can be a totally different colour than it is for us. Other animals are colour-blind between red and green.

The atomists' cosmos was unusual in being neither created nor teleological: that is, there was no overriding metaphysical principle or tendency of things to come into existence in viable ways. In a teleological atomic cosmos, atoms would naturally form complexes of the kind we see around us because this kind of functioning world (or a future progression of it) is the cosmic end goal, just as – for Aristotle – the end goal of an acorn is to be an adult oak. In a more creationist version of this, divine natural law or a god directly brings about such a universe. The Stoic philosophers, for instance, envisaged a cosmos in which god was the acorn, and the oak, and the way in which the acorn becomes the oak, and the reason for it becoming the oak: an ontology which also applied to the cosmos as a whole, which the Stoics defined as a living, divine, material being.

But without teleology or a creator, complicated things happen only if the deterministic atomic laws of weight and motion happen to produce that particular series of combinatory collisions. A complex of atoms, surviving as a temporary arrangement before collisions and internal motions eventually produce its disintegration (death), is not planned or made. It is an accident contingent upon the motion of solid bodies through space. Their direction, speed, and weight (we would say mass, when the atoms are moving in a void, but that distinction was beyond Greek physics) are the only factors determining which atoms bump into which other atoms and at what speeds; and the shape and speed of the atoms involved determine whether they stick together or bounce off each other. Most collisions will not be of the right kind. The formation of complex macroscopic entities like planets and dogs will only result from a very small subset of all the collisions that happen, so if such atomic structures are going to exist in any numbers, there must be a very great number of atoms and collisions in the first place, most of which will come to nothing. Democritus envisaged a world that was full of travelling, colliding atoms, too small to be seen but as thick or thicker in the air (an air made of atoms) as dust particles. Not only his world but the whole cosmos was full of such atoms, and the cosmos was infinitely large. In an infinite space, anything that can possibly happen does happen and, moreover, does so an infinite number of times.

This aspect of Democritus' model, developed as a means of explaining how a cosmos naked of anything but atoms could create the perceptible world, was a good reason for another conclusion: the existence of other worlds within the cosmos. Constrained only by atomic laws of motion, these could take multiple forms:

> [Democritus said that] there are infinitely many worlds, differing in size; that in some there is no sun or moon, in some a sun and moon bigger than in our world, in some a larger number of them [...] and that there are some worlds devoid of living creatures or plants or any moisture.¹¹

Yet others of these worlds would be very similar to each other, and still others identical to the point of having identical people on each world. By this he meant not just that such worlds would, like ours, include humans, but that someone precisely like Democritus would live and think somewhere that was precisely like Greece, and so on for the rest of human history. Conceptually, Democritus has imagined a version of the multiverse scenario developed – for rather different reasons to do with quantum indeterminacy – as a solution to conundrums in twentieth- and twenty-first-century physics. On this hypothesis, infinity similarly allows all physical possibilities to be true somewhere in the multiverse, including variations in apparently arbitrary elements of a universe such as the cosmological constant.

In a Democritean many-worlds universe, atoms are the only constant. They are indestructible, probably because they are composed of completely solid matter with no interspersed space. Atoms in combination eventually break up. A consequence of this is that nothing except atoms lasts forever, not even diamonds or the world or a person's soul. It also means that atoms, arguably, fulfill Parmenides' criteria for something real: they constitute, complementary with spatial nothingness, the unchanging and eternal reality of the universe. Human existence is a real but temporary epiphenomenon: a shadow-play created by the movements of the permanent cast.

Since they cannot have come out of nothing, they have always been there. The atomic cosmos is infinite in time as well as space. (Explicitly so for the Epicureans and possibly also for Democritus, if he considered the question at all.) This point helps strengthen atomism as an answer to Parmenidean problems; it follows naturally from what atoms are thought to be; and it removes the always awkward issue of how to account for a genesis event.

Finally, defining the cosmos as infinite reduces the explanatory need for teleology. It allows there to be many different forms of something, rather than inviting the question of why and how our single world managed to exist in just the way it does. This is a move that takes some of the sting out of the anthropic principle. And because, in an infinite set of possible events, the question of probability becomes meaningless (because everything possible will occur an infinite number of times), the hypothesis of complex events arising by chance rather than necessity or planning ceases to be an unlikely one.[12]

As for the universe being infinite in time as well as space, the Epicureans accomplish this primarily negatively, by arguing against the notion of genesis as a viable option. They take such a genesis to be a creation event, and thus they attack the idea of an origin for the cosmos by attacking the notion of creation by a god.

In antiquity religious and naturalistic philosophy are not different subjects. It would have seemed odd to Aristotle, Democritus, or Galen that 'physics' could be considered a separate subject from 'theology' (using our terms). The question of whether or not the cosmos had a beginning could not be decided without establishing whether creation by a god was a logical possibility and that, in turn, meant debating what a god could and would reasonably be and do.

The Epicureans, for instance, argued that gods were perfect beings (perfect atomic structures) and thereby necessarily self-sufficient.[13] They are eternally disinterested beings with not a care in the world; certainly no desire to design or run said world. Another philosophical school of thought, that of the Stoics, argued on the contrary that the world's arrangements clearly demonstrated a divine providence and that the world, from its beginning to its end, represented the perfect fulfilment of divine identity. In the second century AD the physician-philosopher Galen interpreted the fit-for-purpose nature of human and animal anatomy as proof of the world's teleological structure (as in Aristotle) and also as the work of a beneficent god; but he regarded a god's aims and means as constrained by what was naturally desirable and possible.

This lack of separation between what seem to us to be different subjects with different standards or kinds of argument and proof, had always been a feature of Greco-Roman thought and from the Pre-Socratics onwards explicitly so. As information and expertise in certain subjects developed, specialisation became more common – mathematics is a notable example – but countering this centrifugal tendency was another development: the philosophical 'school'.

The philosophical school or sect (depending on how you translate the Greek word *hairesis*) originated with the centres of discussion for the like-minded founded by Plato and Aristotle as the Academy and the Lyceum. They were a notable feature of Greco-Roman intellectual society from the third century BC to around the fourth century AD. A group of philosophers, beginning as followers of an original and charismatic founder, developed an integrated set of views on the physical nature of the universe, the constitution of a moral or valuable life, the viability and nature of knowledge, and almost any other possible subject. Leading examples who have already been mentioned are the Epicureans (followers of Epicurus) and the Stoics (i.e. 'the philosophers from the portico', an Athenian portico or *stoa* being the place where the friends and followers of the group's founder Zeno of Citium began to meet around 300 BC).[14] A philosophy became a general theory in which physics justified ethics, epistemology defined theology and theology explained physics.

Modern cosmology is more careful about its boxes. The implications of the big bang model for the existence and creative role of god and vice versa have been discussed at length. The continuing conceptual difficulty of something coming to be out of nothing has caused some to propose that a divine role in genesis can be pushed back to the creation of the universe in the big bang, rather than the creation of the world in seven days. This is updating the Judeo-Christian biblical account both to conform with, and to gain the epistemological support of, modern science. Non-believers have responded by arguing that a cosmos coming into being without apparent reason is no more unlikely than a god who likewise exists uncaused. But the choice of cosmologists is between theories derived from the contemporary state of knowledge in physics and astronomy and that are evaluated by a much more limited set of criteria than those employed in antiquity. Modern cosmologists, at least as cosmologists and not individuals with private beliefs, do not attempt to decide between the big bang and continuous creation by asking if a god can reasonably be supposed to be interested in creating a very hot plasma field.

This is partly because we no longer consider questions about gods to fit the same jigsaw as questions about science. Theology lacks equations to solve or phenomena to predict, so introducing theology into modern cosmology would only prevent us from achieving the kinds of answers science provides, answers which are deliberately limited in their scope. We are perhaps more agnostic than the Epicureans, the Pre-Socratics and others about what we can conclude about god(s) through logic and argument, but these are still common methods of debate on religious questions. The greater difference has been in the development of physics and mathematics, from a mode of argument about the metaphysics of the universe, to a more clearly defined and limited investigation of mathematical possibilities and their empirical consequences. This is an investigation which has implications for the metaphysics of the universe, but it does not assume them.

Problems with Time

Greek cosmological philosophy also contains familiar-sounding problems about the nature of time and its relation to a genesis event. In the *Timaeus*,

Plato seems to suggest that time did not exist independently of the cosmos or, as he phrases it (and it is difficult not to), 'before' the cosmos:[15]

> And so he began to think of making a moving image of eternity: at the same time as he brought order to the universe, he would make an eternal image, moving according to number, of eternity remaining in unity. This number, of course, is what we now call time. For before the heavens came to be, there were no days or nights, no months or years. But now, at the same time as he framed the heavens, he devised their coming-to-be. These all are parts of time, and was and will be forms of time that have come to be. Such notions we unthinkingly but incorrectly apply to everlasting being [...] Time, then, came to be together with the universe, so that just as they were begotten together, they might also be undone together.[16]

Plato may well have meant that measurable time came into being with this celestial clock and the divine craftsman's creation, but also that some kind of temporal succession was present within chaos, although no definable present allowed any finite amount of time to pass.[17]

The Roman politician and man of letters Marcus Tullius Cicero (106–43 BC) put it this way, in a fictional dialogue between contemporary proponents of several philosophical schools of thought.

> For if there was no world, it does not follow that there were no centuries. By 'centuries' here I don't mean the ones made by the number of days and nights as a result of the annual orbits. Those, I concede, could not have been produced without the world's rotation. But there has been a certain eternity from infinite time past, which was not measured by any bounding of times, but whose extent can be understood, because it is unthinkable that there should have been some time at which there was no time.
> (*On the Nature of the Gods* 1.21)

Time is still difficult. In many equations of physics, including those that describe physical behaviour at the atomic and sub-atomic levels, those that describe the big bang, and those of the Einsteinian universe in which space–time is a single four-dimensional entity, the universe is time-symmetric.[18] That is, one direction (back, forward) is just the same as another. In the universe that emerges from these equations, the position of a photon at a time we would call the future, some time that has not yet occurred, is just as real and knowable as its position in the past and present; so the future is no less determined than the past. Their reality is of the same status, just as a place to the west of my geographical position in space is no less determined and real than places to my east.

But this is certainly not how the macroscopic scale seems to work, since for us space appears fundamentally different from time. We remember the past, but not the future. We can, or so it seems to us, make decisions that alter the future – I just moved a pen on my desk, altering its position – but cannot do the same for the past. I can undo the movement of the pen spatially, by moving it back, but I cannot alter where it was at 11.44 a.m. Greenwich Mean Time on 24 March Anno Domini 2014. Future events do not seem to cause past ones. And some large-scale laws of physics also imply that time is indeed not symmetric. According to the second law of thermodynamics, in a closed system such as the universe disorder can only increase, not decrease. So on at least the macroscopic level there seems to be an 'arrow of time' – it goes only one way.

Something is lacking in our understanding of how the time-symmetric equations of atomic physics become human perception of the arrow of time. Most people are generally reluctant to take what might be called the Parmenides approach and throw out our experience and intuitions of causality at the macroscopic in favour of a reality in which time really is just like space and the future meets the same conditions as the past, being a matter of perspective only. But what a working theory of asymmetric time would look like remains mysterious.

Many of our intuitions about time – for example, that it exists or that the past is not the same kind of thing as the future – are similar to those of antiquity. The new element is the behaviour of time in the mathematics of the physical world, which is often at odds with those intuitions. Without a clear answer in the same language as to how this applies to the macroscopic and thermodynamic levels, and to our psychological experience, we think about time with the same conceptual logic and attempts at definition as the philosophers of the Greco-Roman era.

Growing Seeds and Bouncing Universes

As we have seen, it is easy to think of Democritus' many worlds as a precursor, not only to the modern idea of the cosmos, full of all kinds of planets and often (in science fiction at least) inhabited by beings remarkably similar to ourselves; but even to the multiverse theory of many worlds or universes. Even the big bang model itself has some parallels with several ancient accounts of the cosmos' beginning.

According to modern cosmology, in the beginning the universe was very very small and very very dense and relatively homogenous. The Pre-Socratic Anaximander of Miletus described cosmogony in terms of a seed's evolution: that is, as something small, dense, and similar all through, but which germinates as a cosmos in fully differentiated flower. Modern popular writing on the big bang frequently uses similar metaphors. Looking for examples I found two in two tries: Peter Coles' *A Very Short Introduction to Cosmology* and the website 'How Stuff Works'.

Anaximander, however, was arguing that the cosmos was indeed an actual living organism, a notion later picked up on by the Stoics. Greco-Roman thinkers often extrapolated from the microcosm (life on the mundane scale of the terrestrial world) to the macrocosm (the universe itself), modelling ideas about the latter on the basis of observations of the former. Such reasoning by analogy was a relatively common move among theorists on a variety of subjects in antiquity: Aristotle's craftsman analogy for nature is only the broadest example.

It was a necessary method of dealing with cosmological questions or any other subject which did not readily admit of direct investigation: hence the common explanation for the nature of the bright celestial objects as fire, or fire mixed with air (*aether*), or, in the case of the Pre-Socratic Anaxagoras' concept of the sun, 'a hot stone'. Indeed, in the presumption that light goes with heat, this analogical reasoning is to a degree correct, but it is also somewhat limiting, since it has to assume that the same conditions apply more or less to the celestial as to the terrestrial, or on various other scales. This is often not the case, and in the instance of the big bang it is notable that modern science argues precisely the reverse.

From a logical point of view, such as those developed by Greek thinkers in response to the Pre-Socratics problematising questions of knowledge, the idea of a singularity, as in the big bang, looks like special pleading. It would seem to raise as many problems about its own existence and nature as it solves about the universe. The Greek mode of thought identified by historical tradition as a precursor to modern science, because of its assumption that things are susceptible to explanation and its emphasis on consistency and proof in such explanations, would have ruled out some of modern science's more unexpected findings – for instance that light is both a particle and a wave – precisely because they appear, on the face of it, to make no sense at all. Such findings are accepted now not only because they have solid empirical or mathematical evidence behind them, but also because experiment and mathematics have proven to be successful techniques in the production of reliable theory.

Sometimes, an apparent correspondence between ancient and modern ideas serves to highlight the very different ways in which those ideas are used or worked out. One example of this is ideas about some kind of cyclical recurrence of the cosmos as a whole.

In contemporary cosmology, it is still unclear whether the universe will expand forever, with its rate of expansion slowing over infinity, or whether it contains enough matter for gravitational pull to eventually match and overwhelm the energetic momentum of the big bang. In the latter case, the universe will contract again as all matter is eventually condensed into a superdense black hole – another singularity. This model is known as the 'big crunch' or alternatively as the 'big bounce', since the recreation of a singularity raises the possibility that a universe in expansion could again emerge from the unique initial conditions of a supermassive black hole. But whether the universe is open (continuing for infinity) or closed and finite, is still a live question because it depends on the ratio of matter to energy and it is not clear how much of either is out there. Estimates of visible matter suggest there is not enough for the big crunch, but the hypothesised existence of dark (i.e. imperceptible) matter and dark energy complicate these calculations. Enough of the former and the universe will contract; but dark energy – which is one explanation for why the universe seems at the moment to be expanding further and farther than visible energy allows for – might be enough to keep the universe open.

Some versions of ancient cosmology offer the bounce without the crunch. According to the followers of the Stoa philosophical school, or Stoics, the world ultimately ends in fire. This is part of the divine plan, and both the world familiar to us and its ultimate conflagration are different instantiations of the Stoics' god, who is a fully material living being, co-extensive with the cosmos (which here consists of the visible heavens plus the earth) and all that it contains. The conflagration occurs when god's plan has been fulfilled: he, or it, then ceases to be the world-cosmos and turns into a fiery sphere. Its fiery form is divinity in its purest form because it is at its least undifferentiated.

This is, however, not a singular apocalypse. The same cosmos plays its history out again with precisely the same individuals and events. The Stoics were early and literal believers in it being impossible to have too much of a good thing. It is precisely because god's plan is good that it does not happen once, but is repeated, exactly, in the same perfect way, every time.[19]

About 150 years earlier, the Pre-Socratic Empedocles (c.490–30 BC) had articulated an elaborate re-occurring cycle of creation poised between two countervailing sources: love, which pulls things together, and strife, which pulls them apart. These two forces gain and lose power so that as one is at its peak, the other is at its minimum. When strife is dominant and love is minimal, each of the four material elements, earth, air, fire and water, that are the cosmos' fundamental constituents, is entirely separated from the others. Life emerges during the period of the cycle when things are neither entirely unified or entirely apart. In its most perfect state, when love is dominant (personified as the Greek goddess of attraction Aphrodite) the cosmos is a homogenous sphere in which the elements and everything else are fully blended with each other. Being thus together as one, they cease to exist as entities in their own right:

> now through Love all coming together into one, now again each carried apart by the hatred of Strife. So insofar as they have learned to grow one from many, and again as the one grows apart grow many; thus far do they come into being and have no stable life, but insofar as they never cease their continual interchange, thus far they exist always changeless in the cycle.[20]

The similarities between the hypothesis of the big bounce and the cycle of destruction and rebirth for the ancient universe are obvious. Empedocles' model also includes two countervailing forces of attraction and repulsion, producing a parallel with the expanding force of initial energy and the countervailing drag of gravity in the 'big crunch'. Other ideas about what the future may hold for the end of universe, and its implications for the beginnings of universes, strike similar themes.

One of these suggests that the universe will continue to expand for billions of years until all matter and energy is stretched beyond the possibilities of classical physics and the universe fragments into multiple versions. Dark energy within these fragments could then convert into matter and radiation and begin another expansive 'bang'. Again, the ways in which we conceptualise the physics of time and space, even down to our metaphors of explosions, stretching, budding and germinations, have not changed much. The same difficulties in thinking of a universe as just popping into being also remain. This partly accounts for the interest in establishing a framework in which such an event is explicable in terms of 'previous' histories and parent universes.

CHAPTER II

STRANGE CREATURES

The Origins of Life

Much Pre-Socratic speculation about the cosmos involved the idea of a cycle. This might be a cycle of destruction and renewal of the cosmos itself, as we have just seen. Or it might be a more limited narrative centred on the world, explaining its origins, development and ending. Unless the world or cosmos was temporally infinite and had always been the same, such accounts needed to explain not just the beginning of the world, but the origins of its constituents and inhabitants.

Anaximander of Miletus, around the middle of the sixth century BC, came up with the first known explanation of both life in general and human origins that seems not to have involved creation by a god. As always with the Pre-Socratics, his own words survive only in the brief reports by other and much later authors. Many of the details are consequently obscure.

According to Anaximander, the first living things were generated out of moisture. They were initially enclosed in some kind of thorny, prickly bark, but as 'they came forth onto the drier part', perhaps as they moved onto drier land or as their environment dried out, the bark broke off. The creatures inside then lived what was described as a 'short and different kind of life'.[1]

A similar kind of process lies behind another report of Anaximander's theories, in which fish or creatures 'very like fish' are produced by the combination of heated moisture and earth. These fish-creatures are necessary for humans to exist: 'In these men grew, in the form of embryos retained until puberty; then at last the fish-like creatures burst and men and women who were already able to nourish themselves stepped forth.'²

It is not clear, from this fragmentary account, how or why humans came to be growing inside these fish-like creatures. There is a similarity in how Anaximander describes them and how he imagines the first origins of life, with the creatures inside their prickly bark. In both cases some form of protective covering is required at this early stage in living existence. In addition Anaximander uses the same language and concepts while discussing the celestial cosmos. His account of the beginning of the world involves fire cloaking the air around the earth, like 'bark around a tree'. This conglomeration of fire later bursts, fragmenting into circles of fire around the earth but separated and closed off by mist: the glimpses we see of these circles through gaps in the mist are the sun, moon and stars.

Perhaps Anaximander thought of the world and its animal forms as having each a kind of life cycle not only for every individual but for each kind of thing as a whole. On the model of individual living things, each such species' lifecycle would involve a birth, followed by a stage of being relatively unformed and incapable of survival, necessitating some kind of protection during their infancy. This reasoning is explicitly attested for his account of human beginnings:

> Further he says that in the beginning man was born from creatures of a different kind; because other creatures are soon self-supporting, but man alone needs prolonged nursing. For this reason he would not have survived if this had been his original form.³

Anaximander's theories about human origins and the beginning of life seem to have been an attempt to solve a specific problem. It was a problem particularly relevant to humans because of their unusually prolonged infancy, but he may have understood it as an issue systemic to organisms and requiring a systemic answer.

What perhaps strikes us is not so much the details of the fish-embryo story but the fact that such a suggestion was made at all. Not only is

Anaximander eliminating gods as a causal factor, but explaining humans in the same biological terms as other forms of life. The idea of human infants and children as enclosed within another animal – possibly suggesting that they were in a sense fish, or reared by fish, is also surprising. The theory of evolution encountered, and still encounters, intense resistance to the notion that humans are related to animals. The inaccurate notion of a biological hierarchy, with humans at the top, is even more widespread. Nor is this solely a modern idea: Aristotle assumed human superiority even as he also argued for animals' anatomy and physiology as beautifully fitted to their function and needs, revealing that contemporary opinion often despised the differently alive or the unaesthetic. In both ancient and modern contexts, then, Anaximander's theory seems remarkably free-form and unrestrained by everyday assumptions in the way it breaks down human–animal barriers.

This was a characteristic he shared with other Pre-Socratics, some of whom had similarly outré theories of human or animal origins. They had created a genre of thought and writing in which radicalism was the only norm and a kind of competitive originality flourished: to a degree, the more startling an idea was, the more memorable and better.

The invocation of fish-like creatures as predating humans probably originated with Anaximander's overarching theory about the world's cyclical transformation from wet to dry, that is from a world consisting of water to one in which the seas gradually receded and the land dried out. In context, then, the choice of 'fish-like creatures' for humanity's incubators makes sense in the general system of Anaximander's theories.

In many ways Anaximander's greatest feat of influence or anticipation was his identification of heat and moisture, when acting on the material of earth, as productive of life. The idea of such spontaneous generation had empirical support, as many commentators have pointed out. Small insect-like creatures such as sand flies appear in numbers in hot, often moist places, seemingly in too abrupt and localised a manner for normal reproduction and with no obvious sexual differentiation.

Aristotle found the idea convincing, endorsing it in his *Generation of Animals* and *History of Animals* for insects, born from decaying earth or plants, and in some cases for living things born within animals from 'secretions of their various organs'. This is evidently a reference to the maggots, beetles or other living things to be found in putrefying flesh, as well as – shades of Anaximander – some species of fish. The ease and ubiquity of such observations, supported by the authority of Aristotle and his successors in natural philosophy, meant that spontaneous generation was a generally accepted part of biology even in the nineteenth century. At that point Louis Pasteur, improving upon earlier experimental attempts to prove or disprove that only heat and moisture were responsible, demonstrated that life did not emerge of its own accord and that micro-organisms in air were a more likely source for even its most mysterious appearances.

The idea of spontaneous generation made good sense in antiquity, not just because it seemed a matter of observation, but because it fitted in with contemporaneous theories of matter: what in the modern world would be called physics, chemistry and biochemistry. Biological creatures had to be constituted from the same kind of materials as everything else, albeit in altered or developed forms. The atomists' basic materials were the most counter-intuitive, but even in that case, a dog and a moon were essentially the same stuff: just a matter of different (atomic) shapes put together in different ways. Other Pre-Socratics reduced the visible world to a purer version of what appeared to be its most vital constituent or constituents. These were usually one or more of water, air, fire and earth. As a quartet, these were given canonical status by the Pre-Socratic Empedocles and then by Aristotle (who added a fifth element, fiery air or aether, for the celestial regions only).

Patterns of change within the world, such as the cycle of the seasons or the varied but limited changes in the weather, were also frequently analysed in terms of an interchange of such basic stuffs and their associated characteristics. Such theories derived from empirical observation of how physical material could transform itself in the right conditions. Heat water and it becomes a kind of air. Air condenses into watery clouds.

In an important way, then, animals, insects and humans were made out of the same materials as the apparently inanimate world around them. Life is, implicitly, what happens when a suitable arrangement of materials is achieved or when they are in the right condition. Empedocles articulated an influential version of the former: all substances that are more complex than his basic four elements of earth, fire, air and water still consist solely of these four, more or less well-blended together, but whether the result is grass or bone depends on the ratio between the elements in that particular substance. The perfect ratio – that is, an even one – between all four elements is only found in blood. As a result, in Empedoclean physiology, blood has a particular emergent property: the capacity of thought and perception. Other Pre-Socratics similarly did not make a distinction between substances that had life and those that did not, given this underlying unity of elemental material.

That was probably the idea behind a view attributed by Aristotle to Thales: that 'mind or soul [*psyche*] is in all things'.[4] This is the concept of 'animism', in which life or even a kind of awareness is a property of everything in the cosmos according to its own nature. Just as existence as a turtle is different from existence from a human, but still a matter of life, perception, and awareness, so is the existence of a rock different from a human as a matter of degree, not as an absolute disjuncture between the living and the non-living.

So, even though individual theories about matter differed considerably, within any such theory there was a continuity of substance between the animate and the even more animate. The additional complexity belonging to blood or flesh was not much greater than the complexity of granite or leaf when compared to the basic element or to the atoms that formed their constituent material.

In general terms, this principle, in which everything consists of the same elements, in different combinations, arrangements and orders, also applies to modern science. Humans, water and asteroids all consist of hydrogen atoms and other chemical elements. But the modern story of matter has many more layers of explanation than the ancient accounts do, both in terms of sheer numbers of elements, and more importantly in the complexity of how they form other material structures, from amino acids to the molecular factory of a cell. This massive increase in the *quantitative* level of difference between complex substances and atomic constitutive parts creates a much greater sense of their *qualitative* differences. In addition, the development of evolutionary and genetic theory has defined life in very specific terms as DNA and RNA, the long chain of molecules which determine, in their physical-chemical structure, the proteins made by living cells.

The spontaneous production of life under the right environmental circumstances is then an easier fit for the imagination of an age before chemistry and molecular biology. Warmth, earth and moisture were imagined as combining their characteristics in quasi-chemical reaction: together, they made a living insect. This kind of generation bypassed reproduction and ideas of inheritance from parent to child, so that such insects and maggots were new creations each time: crawling or flying conglomerations of basic matter.

Implications of Fossils

The conditions thought to be responsible for life were those operative in conditions where such mysterious explosions of life tended to happen – decaying bodies, as Aristotle thought, and perhaps also marshes or hot sand. But they were also conditions associated with life more generally, based on the observations that most living creatures require air (or water, in the case of fish) to survive for any length of time and, again, that animals tend to be warm; observations that similarly drove much medical and biological thinking.

Water and a degree of warmth are still necessary conditions for life, at least life as we know it, to exist. Although exobiology could conceivably involve drastically different chemical conditions, the starting assumption of science is to search for water on Mars and assume that Venus is too hostile. In imagining how life might have begun in the first place, Charles Darwin suggested a 'warm little pond with all sorts of ammonia and phosphoric salts, light, heat, electricity, & c., present', utilising a similar set of conceptualisations about the necessary conditions for life and the possibility of its spontaneous generation.[5]

Darwin was proposing a hypothesis about the origin of all life: an ultimate ancestor from which individual organisms subsequently descended, diversifying into different species. His theory of natural selection was one of several different theories of evolution – of species changing over time – that were developed in the eighteenth and nineteenth centuries, contradicting the majority view that species had been static and separate since their creation. Developments in geological thinking that suggested a much longer history of the earth than Christian theology traditionally accepted, supported evolutionary theory by supplying a timescale large enough for very slow change to be very significant change, through incrementally additional effects. Darwin had set off on the voyage of *The Beagle* with a copy of Charles Lyell's revolutionary *Principles of Geology* (1830–3).

This new geology was founded on multiple observations of the characteristics of rocks in widely dispersed sites. Put together, this information was most plausibly and consistently explained by movements and change amongst the Earth's minerals, necessarily taking large amounts of time. This was a deduction already made in early classical Greece, when the Pre-Socratic thinker Xenophanes of Colophon (*c*.570–475 BC) noted the presence of fossils, often of marine life, in inland locations and argued that this demonstrated that terrestrial parts of the world had previously been part of the sea. The second- or third-century AD writer Herodian reported Xenophanes' views as follows:

> Xenophanes thinks that a mixture of the earth with the sea is going on, and that in time the earth is dissolved by what is wet. He says that he has proofs of the following kind: shells are found inland and in the mountains, and in the quarries in Syracuse [a region in Sicily] he says that an impression of a fish and of seaweed has been found, while an impression of a bay-leaf was found in Paros [an Aegean island] in the depth of the rock, and in Malta flat shapes of all marine objects.

These, he says, were produced when everything was long ago covered with mud, and the impression was dried in the mud. All mankind is destroyed whenever the earth is carried down into the sea and becomes mud; then there is another beginning of genesis, and this foundation happens for all the worlds.[6]

The accumulation of (apparently) eyewitness anecdotal report is impressive in its range and detail: if Herodian's paraphrase reflects anything of Xenophanes' original argumentative strategy, it suggests that the Pre-Socratic utilised the number and geographical extent of such findings to make his point. These were given extra credibility by means of the specifics as to what 'impressions in the mud' had been left as the mud dried out into earth, these being largely marine creatures. Xenophanes' supposition as to the explanation is moreover correct: marine fossils appear in inland rocks, even mountains, because these used to be part of the sea-bed. They are the evidentiary traces of a long-past age.

But Xenophanes produced his conclusions in a very different context to that of the Victorian age. In the Pre-Socratic milieu, radical and startling ideas differentiated one thinker from each other, producing a kind of arms war in innovative and sweeping theories. Moreover, the absence of doctrine about such matters as the nature or extent of the world created a much larger space for such free thinking than did the Christian canon, which offered an authoritative account of creation and the primacy of man that even the authors of the geology revolution, including Lyell himself and Adam Sedgwick, were reluctant to overturn.[7]

It is also clear that Xenophanes' interest in what we would call fossils and his conclusions about what they say about the earlier history of the earth are framed in terms of the familiar Pre-Socratic questions of how the world as we know it came to be and what will ultimately happen to it. He seems to have argued that the terrestrial earth has been and is being produced as the fundamental elements of the world transform from one into the other, the wet into the dry; and it will be destroyed by the other half of the same cyclical process, as the dry eventually becomes the wet –

ocean – once more. Again, then, we can see how a recognisably 'scientific' process of gathering and interpreting data is only part of the story. Xenophanes introduces fossils as evidence for a much more general – and less well substantiated – account about beginnings and endings.

Origins of Species

As for an evolutionary theory of life, there was little need for one in antiquity. Aristotle's species were formed, and had always been so, by the needs and constraints of their nature according to the internal logic of the cosmos. The question of the origins of life, animals and humans did arise, but as a problem involving a single-generation beginning rather than a continuing process, as in Anaximander's postulated genesis for bark-creatures or human exogenesis. The question of whether species have changed was not posed. The particular characteristics of any species or genus (to use terminology devised by Aristotle as a classification of functional relatedness) were considered in terms of whether they could be explained by larger principles of explanation, such as teleology, design or accident. Darwinian evolution is an answer to a puzzle about the origins of individual species, or rather as to how species might change and change so far that they become something else. Ancient enquiries viewed the origin of living things as a single problem with a single answer, one that indeed was not fundamentally different from explaining the origins of non-living things.

Atomism, in its rejection of design and teleology, might seem the most likely candidate for ideas of evolution. Individuals must have come into being as accidental atomic conglomerations. In theory, one could write an atomic story approximating to natural selection, since only some atomic complexes would succeed in sticking together as organisms and only the right kind of organic complex would happen to have an atomic organisation whose physics necessarily led to their consistent self-replication (reproduction).

The Epicureans, in fact, do offer an account of the early period of life on earth that contains a role for an idea very similar to that of natural selection. They seem to have adapted this theory, however, not from Democritus' atomism, but from a different Pre-Socratic. I discuss the later Epicurean version in more detail below, but its original was an aspect of Empedocles' cosmic cycle (Chapter I, 'Growing Seeds and Bouncing Universes'). As love increased and strife decreased (and vice versa), the four fundamental elements were attracted to each other, forming larger and more complex mixtures, including whole objects with differentiated structures. An arm, for example, the flesh and bone of which were constituted by the four elements but in different ratios of each.[8] As love continued to increase, these kind of structures continued to combine, this time with each other:

> Many creatures were born with faces and breasts on both sides, man-faced ox-progeny, while others again sprang forth as ox-headed children of man, creatures compounded partly of male, partly of the nature of female, and fitted with shadowy parts.[9]

Another paraphrase of his argument describes how these strange creatures were modified, not by a designer but by their accidental ability at living:

> Wherever, then, everything turned out as if would have if it were happening for a purpose, there the creatures survived, being accidentally compounded in a suitable way; but where this did not happen, the creatures perished and are perishing still, as Empedocles says of his 'man-faced ox-progeny'.[10]

The crucial point of similarity between this and Darwin's theory of evolution by so-called 'natural selection'[11] is not only the absence of teleology or design, but the observation that out of a multiplicity of accidentally generated, chance combinations, some would survive and others would not. Empedocles' account is still an all-or-nothing affair for individuals, apparently within a single generation; whereas Darwinian selection is a matter of relative survival rates over generations, but the idea of non-arbitrary survival as a result of attributes acquired by chance is the same in both cases.

Empedocles' theory is reminiscent of the notion of 'hopeful monsters', a phrase coined by the early-twentieth-century geneticist Richard Goldschmidt as part of his argument that small accumulative changes in the genome, of the kind generally described by Darwinian evolution, were insufficient to explain speciation.[12] Mutations in important developmental genes could have major effects that produced a kind of animal very different from its predecessors: so different it was a 'monster', but one that might be capable of being a successful new species: a 'hopeful monster'.[13]

This is one version of saltationist theories of evolution, in which speciation or other change happens in abrupt, obvious 'leaps' of macromutation. Saltationism has died out in modern evolutionary theory but had some notable adherents in the nineteenth and early twentieth centuries: thinkers to whom differences between species seemed too wide to be explained by anything other than a great leap forward. Many twenty-first-century creationists make a similar argument, arguing that the diversity and complexity of life is too great to be explicable as the effects of small differences in genetic code.

The alternative story of speciation, incremental change, relies on both long periods of time and, often, the eventual death of intermediate forms as less successful than the more fully realised versions. It is not clear how long a period of time Empedocles allotted to the cosmic cycle as a whole. Possibly 1,200,000 years was the period between each occurrence of the perfect sphere in which everything becomes a single completely blended substance.[14] Fragments imply that 10,000 years was the total duration of either the period in which love pulled together disparate elements, eventually into viable organisms, or in which increasing strife made mixtures separable, with the same effect. How long, within this overall length of time, each of the various stages of the emergence of life-forms took is unclear, but presumably much of it involved a slow process of either attraction and synthesis (as love strengthened) or repulsion and differentiation (as strife strengthened) into distinguishable parts and combinations of parts. It was the first generation that took the time.

If we stretch this analysis even further, continuing to read Empedocles' origin story as a biological theory, we could conceivably argue that his unsuccessful combinations-of-parts are the ancient equivalent of extinct species, in particular extinct intermediate forms which were simply not good enough at being either men or oxen, but were an unsatisfactory combination of the two.

The difference would be that of saltationism to the incremental evolution of contemporary theory. As in saltationist theories, in the Empedoclean account, the failure of a particular combination of parts in a single individual is the same as the failure of a would-be species. Some monsters are hopeless. But in Darwin's theory of evolution, species evolve not as a single monstrous individual, but through the differential survival of slightly variant individuals (Darwinian evolution) or through the differential survival of genetic elements within a population (an aspect of the neo-Darwinian synthesis popularised by Richard Dawkins' classic *The Selfish Gene*).[15] In saltationism, the failure is instant. The gap between an old successful species and a new successful species has to be crossed totally, or the hopeful monster simply falls into the abyss – that is, its new characteristics make it unable to survive and/or reproduce.

In Darwinian evolution, although individuals may fail to survive, intermediary forms of species flourished for at least some time (and some, like lungfish, which are an intermediary form between fish and air-breathing organisms, are still with us). The jump or gap appears sizeable principally because these intermediate species and individuals have now died out or become relatively rare, because even more successful forms eventually out-competed them, so to us it often seems as though a single species transformed into something significantly different without any intervening steps. But this is an accident of historical perspective and gaps in the fossil record. For example, in Darwin's day there seemed to be no intermediate species between dinosaurs and birds, but *Archaeopteryx* fossils, of which the first discovery was published in 1861, have characteristics of both dinosaurs and birds (feathers), and are an early example of a transitional species.[16]

Empedocles also shows continuity of thought with another evolutionary theory. Like saltationism, the early evolutionary theory now chiefly associated with the French biologist Jean-Baptiste Lamarck and therefore known as Lamarckism has been discarded, but it, or aspects of it, was for some time quite commonly accepted.

Darwin's chief difficulty in working out and convincing others of his theory was the universal ignorance, at the time, of the actual components and mechanism of inheritance in plants and animals.[17] As a result, theories of so-called 'soft inheritance', that is inheritance through acquired characteristics, seemed plausible enough to many and indeed *The Origin of Species*, especially later editions, allows a role for this mechanism.[18] The assumption was that anything that happened to an organism prior to reproduction would or could be transmitted to its offspring. A scar on the parent's arm, for instance, would also appear in his child (or was as likely to do so as any other characteristic).[19]

Lamarck combined this with a belief in the adaptive effects of the environment on an animal. In striving to run faster, or see better in darkness, individual animals improved their own abilities in these regards, just as regular practice makes a human a better piano player or a faster, stronger, athlete. These developed, acquired characteristics could be passed on to its children, where they might either atrophy through lack of use or be developed even further. In this way a line of descent tended to improve at particular skills over time, eventually producing a new species – cheetahs, for instance, or owls.

In literary form, this kind of thinking is represented in literary form by Kipling's *Just So Stories* of how the elephant got its trunk or the leopard its spots. An accident turns out to be a lucky accident and produces a new kind of creature. That Empedocles may have offered a similar kind of origin story for some kinds of animal is suggested by a passing remark of Aristotle, in the context of criticising Empedocles' non-teleological stance:

Genesis is for the sake of what is, not what is for the sake of genesis. Hence Empedocles was wrong in saying that many attributes belong to animals because it happened so in their genesis, for instance that their backbone is such because it happened to get broken by breaking.[20]

We note again the emphasis on accidental changes in a life-form. The example also suggests a belief in acquired characteristics, although not necessarily Lamarck's analysis in which environmental pressures drove the acquisition and development of such characteristics. Empedocles' accidental backbones are more similar to *Just So Stories*. Again, if we construct Empedoclean thought in our terms, as a biological theory, it looks as if he explained dramatic differences in organisms as the result of happy accidents, whereas Aristotle argued that the logic of nature inevitably produced functional structures where necessary. It is unclear how Empedocles related this kind of story to his insight that what was crucial were accidents that worked, or whether he was simply trying to account for differences among what species there were in existence. Caught between the twin poles of teleology and pure chance, he was unaware of the power of his own idea.

Returning to atomism, extant Epicurean materials also invoke a kind of natural selection. Our chief surviving source is the Epicurean poem *On the Nature of Things* by Lucretius (*c*.99–55 BC). He asserts that the early earth was far more fertile than in his day and consequently produced all kinds of life forms by spontaneous generation. Using language very similar to that of Empedocles some 400 years earlier, he goes on to describe how some of these often monstrous forms were quite unfitted to survive to adulthood (*On the Nature of Things* 5.837–58). And although the extant fragments of Empedocles are not explicit on this point, in Lucretius at least the added criterion of the necessity of *reproduction* for a species' survival is emphasised: 'Many animal species must have become extinct at that time, unable to reproduce and spread their progeny' (5.855–6). Only those accidental animals whose speed or power or cunning or usefulness to humans enabled them to succeed at feeding, defence and sexual intercourse survived for the modern world.

Having eliminated monsters, cultural and biological evolution proceeded primarily through exploiting existing features of organic existence and the physical world. Language, for example, was a reuse of vocalisations for identifying and referring to objects. But the vocalisations themselves, it would seem, were themselves an opportunistic use of how some animals were (accidentally) assembled in spontaneous generation. This use was then passed on through reproduction; so Lucretius says that there was no seeing before there were eyes with which to see; and no use of fire before lightning demonstrated its potential.

As we saw earlier with respect to Empedocles, there is a very different explanatory context for, and significant differences between, Epicurean 'natural selection' and modern evolutionary theory. In the first place, Epicurean theory is the explanation of a particular stage in the history of life that immediately follows the spontaneous generation of a differently fertile age (equivalent to Empedocles' double zoogonic period at the moments when love and strife are in the right kind of counterbalance). Since then, species have been entirely or largely static.

Secondly, Lucretius' version suggests that organisms and animals had to learn to use their organs or other abilities, imparting a Lamarckian flavour. But they needed such useful organs in the first place: individuals found themselves either well-endowed or doomed. Perhaps this absence of an idea of incremental change explains why Lucretius does not discuss the origins of modern species as atomic compounds but only in terms of a larger biological scale, starting with the generation of life forms formed, Empedocles-style, from random collections of organs and compounds. If one re-imagines the atomist account, it might seem inevitable that some accidental assemblages would fit together worse than others, so that they might shake apart earlier or replicate less successfully. But the atomists could have answered that it was instead an all-or-nothing question of existence. Either an atomic complex is a successful organism (or non-organic structure) or it is not.

Ancient atomists emphasised accident in specific opposition to arguments from design or teleology. The conditions of infinite time and an infinite supply of atoms were crucial to their case. This very similarity with Darwinian theory, which also required very large amounts of time as well as continuous natural variation between individuals, is a factor in antiquity's divergence from Darwinism. It created a context that was unfavourable for, or even hostile to, natural selection and it did so for precisely the opposite reasons to the preference for design often articulated in the nineteenth and twentieth centuries.

William Paley famously complained that Darwinian evolution was like finding a working watch, in all its intricate clockwork mechanism, randomly assembled by wind or other physical forces out of bits of dirt and stone and metals. The atomists' objection would, I suggest, have been not that Darwinism relied too much on chance – the creationist objection – but that it did not give chance enough credit, and instead placed a much greater emphasis than the atomists did on a mechanism which improved upon chance. They would cheerfully have accepted Paley's image: of course watches, eyes and humans can be assembled simply by the contingent cause-and-effect of physics, so long as you have enough time, space and atomic matter. In antiquity the failure of some life forms to survive is a theory developed to account not for species as they are now, but the absence of other kinds of life – of monsters. As modern writers on evolution sometimes put it, there are many more ways of being dead than being alive.

The Origins of Successful Theories

We know very little about Empedocles' life or about the circumstances in which he developed his theories. He was probably born in about the year 490 BC. He is said to have come from the Greek colony of Acragas (modern Agrigento) in Sicily, that is in the regions west of the Aegean that had been settled by the Greeks over the preceding century and a half. But later biographies, which include the story that he killed himself with a leap into local Mount Etna, are largely apocryphal. He wrote in poetry, the form of speech associated with divine inspiration and also used by several other early Pre-Socratics. A third-century AD source identifies two such works, 'Concerning nature' and 'Purifications': the fragments of his ideas quoted by other authors have therefore traditionally been ascribed to one of these two works. Those that seem concerned with natural philosophy are assigned to the former and those concerning reincarnation or morality to the latter.

But this division, which relies upon the rather anachronistic supposition that Empedocles thought, like us, of religion and 'philosophy' (a term not yet invented) as two separate categories, may be unreliable. It is not clear that there were in fact two poems and an increasing number of scholars have argued that we should regard his work as essentially unitary. Empedocles presented himself as a charismatic, semi-divine figure with the ability to cure the sick, change the weather, ensure the fertility of crops and raise the dead. His radical reconfiguration of traditional ideas about gods, souls and the meaning of life integrated, as part of its internal justification, a new story about the origin and nature of the world, living things and humans, often in considerable detail – he developed, for instance, theories on the physiology of sense-perception. This set of concerns is what strikes us as falling under the category of science, but what Empedocles was offering was a set of claims, part-reasoned, part-inspired, about the moral as well as the material nature of the world and the appropriate way to understand and live in it. Some of his fragments imply an element of personal salvation to this in a progressive reincarnation up a scale of organisms, from plants and then animals to men and eventually *daimons* (minor gods): this last being a status he seems to have claimed for himself. As with other Pre-Socratics and indeed many later philosophers, his arguments for the nature of the world consisted of reasoned speculation. This was compatible with observation but rarely involved sustained investigation, and was as motivated by questions about epistemology, theology, morality and metaphysics as by any physical subject of inquiry.

Charles Darwin accepted the position of the resident naturalist on the Royal Navy mapping expedition of *The Beagle* after an undistinguished undergraduate career that was notable, in retrospect, chiefly for his attention to detail in capturing beetle specimens as a part of a sideline in natural history. Too squeamish for medicine, he was unenthused by his original prospect of becoming a clergyman – a default occupation for many well-educated Englishmen. The work he did during *The Beagle*'s five-year round-the-world expedition, however, provided him with an opportunity to put his collecting skills to extensive use and developed his skills as a naturalist and geologist. His accumulation of information on all kinds of plants, animals and geological landscapes was an enormous tranche of evidence that later proved vital to developing his arguments for the evolutionary mechanism of natural selection. Already familiar with new ideas about geological time and the grandson of an earlier evolutionary theorist, Erasmus Darwin, he was mentally prepared for encounters with a world that the traditional Christian story of unchanging and perpetual species since Creation did not seem to adequately explain.

His unpublished notebooks reveal that, as early as 1836, before *The Beagle* returned to an English port, he was privately convinced that species were not static, but underwent change and evolution. In 1838 he read Thomas Malthus' *Essay on Population* and immediately perceived that its central theme of competition for limited resources could be the force driving change in species by 'thrusting out weaker ones [i.e. weaker adaptive structures]'.[21]

In 1844 he wrote an essay summarising his ideas, to be left in the custody of his wife Emma and handed over to specified potential editors in case of his early death. But he did not start writing *On The Origin of Species* for another 14 years, until August 1858, completing and publishing the manuscript the following year. He had been an evolutionary theorist for some 22 years and in possession of a distinctive insight for roughly 20 of those, but even within the scientific community very few people knew it.

The event which finally prompted publication was the realisation that the naturalist Alfred Russell Wallace had arrived at the same idea. Unaware of Darwin's own theories and still east of Singapore, Wallace sent his brief essay to Darwin as a preliminary to presentation to a wider audience. The story of how this finally prompted Darwin to risk making his own work known outside a very limited network of family and close colleagues has prompted much discussion. Panicked but at first resigned to losing any claim to originality, he and two influential friends rapidly arranged a joint presentation of Wallace's outline and written material of Darwin's to the Linnaean society, and then wrote to explain the situation and their *fait accompli* to the roving Wallace. This ensured that both men shared the credit for the idea, but that Darwin was acknowledged as the first theorist with the most developed ideas. (Both true, but Wallace's idea was arrived at independently and he might have legitimately pre-empted Darwin into print.)

Perhaps his friends were relieved that events had finally nudged Darwin into action. One, Joseph Hooker, kept the momentum going by urging Darwin to write a fuller and up-to-date account of his theory for publication in the Linnaean society's research journal (the version presented to the meeting itself had been the 1844 outline for literary executors, combined with a letter of 1857). Darwin's reaction is revealing: how would he make it sufficiently scientific without 'giving facts' and in a mere 30 pages?[22] What he eventually wrote turned into *The Origin of Species*. It contains many such 'facts' – empirical evidence – and it is much, much longer than 30 pages.

Wallace's willingness to publicise his germ of an idea immediately contrasts sharply with Darwin, who behaved more like a goose that has laid a golden egg and proceeds, alarmed by this attention-getting event, to sit on it for two decades. But his caution had its uses. Recognising just how controversial would be the news that humans are descended from other animals and closely related to apes, Darwin was intent on making his idea as bullet-proof as possible. His work on the specimens and observations derived from *The Beagle*'s voyage took up much of his time for several years, but also represented a wealth of data, such as his records of the flora and fauna of the Galapagos islands, which could now be studied in more depth. He had spent another eight years classifying his specimens of barnacles, publishing the results in two volumes over 1853–4: a project that provided the kinds of 'facts' that in 1858 he thought so vital to proving his theory. Darwin's work on barnacles was an irrefutable taxonomic, anatomical demonstration of the relatedness of one species to another and how minute variations distinguished one individual from another: a set of observations that revealed how extensive was the material on which a selective process might work.

He used experiments to test out possible answers to the flaws and obscurities in his theory. In particular he explored the ways in which new animals and plants might have reached new locations like islands, counting the seeds on the feet of (dead) birds and discovering that asparagus seeds could germinate after 32 days in salt water. His wife's floral hats were only one method by which he investigated the preferences of insects in pollination. Hybridisation became an acute interest, as did the breeding of domestic animals and plants. Correspondence with pigeon breeders, bee-keepers, gardeners and anyone with potentially useful information gathered what turned out to be an enormous amount of supportive evidence. About 14,000 of the letters that he wrote and receive survive, and the editors of this voluminous correspondence estimate that an equivalent number have been lost.[23] Discussion, usually via yet more correspondence with those few notables who did know of his ideas, including botanists like Joseph Hooker and Asa Gray, the anatomical expert Thomas Henry Huxley and even evolutionary sceptics like Lyell, refined and improved his arguments.

The result was that *The Origin of Species* is a long book full of evidence, closely and carefully argued, that disarms objections by discussing, explaining or simply admitting them. It is much longer than the original Wallace–Darwin submission, Wallace's initial presentation or the sketch of the idea Darwin gave to his wife as a precaution against his death. This is not so much because the actual insight is complicated – the theory is one of the few in science which can be explained in its essentials on something as small as the back of a postcard.

What Darwin's 500 pages accomplish, based upon those decades of additional research into biology and botany, is increased persuasiveness. It does this through his accumulation of supportive evidence, parallels from domestic breeding – which readily demonstrates how flexible characteristics are – and exploration of potential difficulties, such as how starter populations of certain species could reach remote islands before they differentiated there to fill every available niche. He also omitted to point out the implications for the origins of humans or god as a creator. Alert readers could draw their own conclusions, but the general tenor of *Origin*, unlike its content, is non-revolutionary. It does not take on opponents or highlight its own radicalism.

It is impossible to say how much difference this made to its reception. Wallace's independent insight demonstrates that the intellectual environment, at least, was not only receptive but productive of evolutionary ideas, not least in the nineteenth-century trend for careful natural history in remote places bulging with life. But it seems reasonable that *Origin*'s accumulation of evidence had some persuasive effect on those who read it. Moreover, much of Darwin's thought on the relevant issues proved remarkably secure: not only in evolution in general but on such questions as sexual selection or his rejection of group selection, a notion emphasised much more by Wallace.

Wallace's bombshell had been timely in two ways: it pushed Darwin into realising that he could lose the credit for having the idea and doing the work, but it had been delayed long enough to allow him to make that idea as credible as it could possibly be at the time, given the lack of knowledge about genetics. His work appeared in a culture which had a more established and consensual epistemological framework for scientific inquiries than was the case in antiquity, and it avoided open conflict with religion by dint of not discussing it.

This was a very different approach to the speculations of charismatic Pre-Socratics who began, not from the empirical mystery of why islands contained such similar but distinct species, but with the questions of how everything began and what this meant for men. Darwin and Wallace's ideas depended on prior work in geology, which greatly extended the available amount of time, and derived from their and others' experiential research in natural history and anatomical taxonomy: already a sizeable body of knowledge at the beginning of the eighteenth century and which increased dramatically over their lifetimes – partly thanks to their work – and over subsequent centuries. It is a notable feature of the post-*Origin* debate over evolution that it often turned on details such as the difference between gorilla and human cranial anatomy, on questions of hybridisation or on geology's reliability in estimating the age of the earth. *On The Origin of Species* helped to make this shift in the terms of discussion, from the philosophical framing of previous works on evolution towards one in which empirical standards were the criteria of persuasion and proof. Moreover, Darwin and Wallace, not to mention their scientific opponents, were part of a well-established scientific culture with a considerable degree of consensus on methodology and an already existing body of knowledge in physics and the natural world. The rapidly burgeoning number of scientific journals and embryonic peer review institutionalised the establishment of consensus even as it provided a managed arena for dissent and novelty. When disagreements occurred, therefore, they took place in a context where people could at least agree on what they were disagreeing about and what kind of evidence would be conclusive one way or the other.

Comparing the context, use, and argumentative framework of Darwinian natural selection with its Greco-Roman predecessors is a good example of how science is embedded in culture and society. Empedocles and the Epicureans had the right idea at the wrong time, with no established means of rendering it noticeably more persuasive than any other idea. They lacked a culture, either general or even intellectual and small-scale, that prioritised empirical evidence. And they made the logical insight that is at the heart of evolutionary theory as part of a speculative biology, geology and physics that was largely a fiction and at best too generalised and imprecise to be of much use. Even Xenophanes' fossil-based observation that the current division of sea and land was not permanent formed part of an assumption of destruction and renewal as a fundamental principle of reality, rather than an interest in geological processes and history. The Pre-Socratics and their successors in Hellenistic philosophy were not particularly interested in what we think of as scientific subjects for their own sake, and as a result, their suggestions coincide with science only on occasion.

Plato and Primates

As an addendum, it is worth noting that the comparison drawn by Darwin himself, between his ideas and those of classical antiquity, involved not Empedocles but Plato. In one of his private notebooks, he remarks: 'Plato [...] says in Phaedo that our "necessary ideas" arise from the preexistence of the soul, are not derivable from experience – read monkeys for preexistence.'[24]

In his dramatic dialogue the *Phaedo*, named after one of its characters and taking place in the hours and minutes before Socrates' execution in 399 BC for religious impiety and corrupting the young with novel ideas, Plato explores the notion that humans can conceptualise such things as justice, equality and beauty because our mind or soul (the Greek word is *psyche*) has previously encountered the true, complete and unambiguous nature of these items in a non-material reality. It is an argument that depends largely on disproving the counter-explanation, subsequently articulated by Aristotle, that these concepts are developed by our experience of empirical everyday examples of them (and of their opposites: untrue, unequal, ugly).

Plato argues that we never encounter an unambiguous example of such concepts. A stick of a length equal to another stick will be unequal to the different length of a third, making the first stick as unequal as it is equal. Helen of Troy is only beautiful compared to humans, not to Aphrodite. Returning an axe borrowed from a neighbour is the right thing to do, but not in all circumstances, such as when that neighbour is threatening to axe-murder his wife. Given this absence of any example of justice, beauty or equality that is reliable regardless of context – any example that is not also an example of injustice, ugliness or inequality – Plato argues, not that circumstances alter cases, but that we could, in such a universe, never have developed concepts applicable to circumstances at all. We would have no epistemological beginning point. The only option left, therefore, is to conclude that the *psyche* has encountered these concepts in some other reality than that of experience, specifically before instantiation within a material body.

It can therefore recognise what is really just or good about returning an borrowed axe in the appropriate circumstances (to a happy and rational owner), without making the mistake that the justice just is 'returning what you have borrowed' regardless.

Darwin's serious joke is to update this notion of an innate understanding of conceptual items. They become not immaterial entities encountered by an individual's eternal soul, but the result of an evolutionary response to environmental pressures that demand systems of evaluation and response. Subsequent research has supported the theory that organisms have evolved to pick out certain aspects of their physical and social environments that are related in particular to survival and reproduction: we do not see everything equally but only things relevant to us: the known individual in the crowd, our name in the hubbub of a party, the quick movement of a snake in the undergrowth. We search for and sometimes over-perceive patterns in the information bombarding us. Monkeys are able to perceive – that is, to understand – when another animal has been given more food than they have: a grasp of numerical inequality and inequity. They can perceive this as unfair: a rudimentary concept of (in)justice. And they presumably have standards, perhaps largely pheromonal, for beauty among monkeys. They have these abilities because evolution has favoured individuals with an incremental advantage in assessing amounts of food, in managing their behaviour in regard to others in a social group, and in finding fertile members of the opposite sex attractive.

The point is more general than the abilities of monkeys. Darwin could have said 'reptiles' with the same meaning, although the cognitive presets would have differed somewhat. The 'necessary ideas', the conceptual forms visible in humans, have been made by evolutionary processes through generations of our ancestors: they are the genetic and phenotypic codification of what has worked in the past as a means of interpreting and managing the world, and therefore of surviving within it, just as much as your optic nerve, your heart and your backbone.

Darwin's jotted analysis is not the simple recasting of a concept in a different language, but the successful supplanting of one theory by another. In that, it can also stand for the failure of Platonic styles of argument to withstand the approach of natural history – a methodological approach descended from that of Plato's one-time student, Aristotle. In the next chapter, we will explore the invention of 'nature' and its empirical study as a Greek conceptual category and field of inquiry, and examine Aristotle's influence upon science then and now.

CHAPTER III

NATURAL LAWS AND HOW TO DISCOVER THEM

The Category 'Nature'

The ancient Greek word *physis* comes from related words for the birth and growth of living things. Animals and plants are complicated entities that change over time and interact with their environment. Plants, which begin as small undifferentiated seeds or nuts, develop stems, leaves and sometimes flowers. They grow larger over time. They produce seeds of their own, resulting in new generations. They require air and water if they are not to die and disintegrate. Animals have similar capacities, as Aristotle points out, but in addition can move from place to place, eat, drink and usually reproduce through sex. Fish seem to breathe water, but cannot survive air. Insects, the Greeks thought, often reproduce asexually. Although there are differences, between species, genders and individuals, these complicated processes happen for the most part in a predictable and consistent manner. Kittens become cats, not starfish or humans. The parts of a body grow in proportion to each other. The more an animal eats, the larger it gets, but not beyond a certain size – after that, fat or bulk is added, but no-one becomes a giant. Some things are the same across different species: blood, bone, the male and female sexes.

Physis came to be a way of referring to this underlying regularity. It designated a category of change in which things happened without external intervention, but on a non-random basis. A bolt of lightning might strike an oak sapling dead, but in the normal course of events the sapling would have become an oak tree, with no human action required. This is what happens naturally or, as the Greeks put it, *kata physin*, according to nature'. Although the word itself originated in a biological context, nature is visible in other parts of the world beyond plants and animals, wherever there seems to be structure or regular change that happens by itself. The change of the seasons is an example of this larger natural world. So is the propensity of a rock to fall downwards rather than upwards. In particular, people observed the long-term regularity of astronomical phenomena, from the alternation of day and night to the risings and settings of stars and constellations at particular times within a yearly cycle.

These were not just matters of intellectual interest. Farmers used the fact that the positions of constellations and stars varied with the seasonal time of the year as a guide to what to plant and when. Navigators were similarly reliant upon the map of the night sky, while weather and climate were widely believed to be affected by the movements of the stars and planets. Astronomical events formed a significant part of the category of *physis* at its most invariant and precise. (For more on astronomy and mapping, see Chapter V.)

Physis, then, meant both a category of what usually happens in the world around us; and an explanation for a particular kind of causation, something innate to an entity or system. The emergence of this concept was a large part of the different kind of explanation that was favoured by the Pre-Socratics and beginning to define much of medicine. Explaining an event or entity in terms of its 'nature' depersonalised notions of causality. Instead of a god or magic being responsible for epilepsy, distress, drought, lightning or love affairs, these and almost anything else became the products of factors innate to human physiology and, ultimately, elemental interactions. Explaining the why of a thing shifted from a narrative of external agency – often conceptualised as an attack from enemies or angered gods – to internal mechanisms without intent. In one probable example of this paradigm shift, doctors ceased to explain epidemic disease as something acquired from contact with those already ill. Instead they argued that common environmental factors, such as a change in the wind, were responsible for a single widespread illness (Chapter IV).

NATURAL LAWS AND HOW TO DISCOVER THEM

Nowadays we usually refer to this style of explanation as naturalistic. It offered the Greeks a very large arena for investigating and speculating in new ways about common phenomena and problems, with a resulting explosion in debate and literature. Naturalistic terminology and theories quickly became the dominant mode in technical genres such as mathematics, astronomy and medicine, and were an important element in philosophical inquiry. But they did not transform the culture as a whole. Even among the relatively small group of educated, literate, wealthy individuals at leisure to debate such ideas, there was resistance to the notion that reality could be reduced to an understanding of something's internal structures and mechanisms – a notion that lay at the heart of the concept of nature.

Plato, for instance, characterised his teacher Socrates as having attacked empirical investigation of the world, but others associated Socrates with non-traditional opinions in general. Plato's older contemporary, the satirical dramatist Aristophanes (*c.*450–388 BC), portrayed Socrates in the *Clouds* as an early example of the mad scientist or at best absent-minded professor trope, with Socrates' head 'in the clouds' (the phrase comes from this play) as he invents an instrument for measuring the distance jumped by a flea. Socrates was eventually convicted by an Athenian court in 399 BC of corrupting the young and impiety toward the gods. In the intellectual and political ferment of fifth- and fourth-century BC Athens, attitudes that seemed to question conventional ideas of religion or morality formed part of a culture war in a febrile politico-legal system. Young, politically active aristocrats formed an admiring entourage for Socrates and other radical teachers, the so-called sophists or 'clever men'. These itinerant thinkers-for-hire have been described by some modern writers as mobile university professors, but they have just as much in common with modern charismatic exponents of culturally sensitive religious and political movements. They attracted similar notice and controversy, and in many cases their relationship with their followers was more that of master and disciple, demonstrating commitment to a distinctive way of thought or life, than the educational imparting of widely accepted knowledge.

Since naturalistic explanations downgraded or reframed the traditional roles of the gods, and were closely associated with such radical theorists as the Pre-Socratics and sophists, they were potentially as controversial as Socrates' questioning of conventional moral definitions. Aristophanes' play also suggests that they were seen as deeply impractical and irrelevant, although we should remember that comedic satire always takes the extreme point of view. But other evidence points the same way. An anecdote circulating in this period about the Pre-Socratic Thales of Miletus, to whom Aristophanes had compared Socrates, had him falling down a well because he was looking, in an astronomically minded way, at the stars. Meton of Athens, sometime around 432 BC, worked out a solar calendar far more accurate than the lunar ones used by Greek cities at the time, but there is no evidence that either his home state or any other ever adopted his calendar as part of civic life. (This is discussed in more detail in Chapter V.)

The fifth-century BC historians Herodotus and Thucydides, while plainly acquainted with naturalistic medical theory, also demonstrate a degree of scepticism. Thucydides seems to accept the non-Hippocratic notion of contagion (see Chapter IV), and Herodotus sometimes explicitly sets up a traditional or religious explanation for some event – a defeat, a king's madness – in contrast to a more naturalistic or material cause, and then says which he prefers (usually the former). So, although naturalistic explanation was a well-known new idea, and intellectuals were familiar with its terminology and approach, it was adopted cautiously and piecemeal as it percolated slowly through contemporary culture.

Nature was a highly normative concept in that it established general rules about what normally happened, making exceptions appear aberrational. The concept was also strongly associated with functionality, appropriateness, and biological necessity.[1] In modern parlance, Greeks were happy to move from an 'is' (what normally or naturally occurs) to an 'ought' (what should occur; what is best). Much naturalistic theory, however, created norms on the basis of social preconceptions and limited evidence. For instance, theories on female biology offer some notable examples of how inaccurate ideas of what was natural for women seemed to physically confirm their social roles and identity (see Chapter IV, 'Theories about Women').

Investigating the nature of things (this is the literal title of Lucretius' Latin poem on Epicurean physics, *de rerum natura*) required establishing what does happen 'naturally', that is what happens when no external force (such as a lightning bolt killing a sapling) interferes. Much of this knowledge came from common observations and traditional knowledge, or – as in theories of female biology – was heavily influenced by preconceptions, but overall naturalistic explanation fostered a trend towards empiricism and a search for consistent laws of behaviour. Aristotle took this qualitatively transformational step further, towards an approach that seems often recognisably scientific in its aims and methodology.

Aristotle and the Modern Scientific Method

One could say that Aristotle was the first person to invent the notion of science, although this does require a large dose of hindsight and a generous approach to definition. His division of philosophy into theoretical, poetic and practical goes some way towards establishing a separation between human activity (poetic philosophy, or what we might think of as the arts and the practical philosophy of ethics and politics, possibly somewhat equivalent to the social sciences) and the subject matter of how things worked (theoretical philosophy). The last grouping comprised mathematics, the natural world (*ta physika*) and metaphysics. This, and especially the sub-category of natural philosophy, is roughly equivalent to the domain of enquiry that we would associate with science.

Aristotle argued that knowledge of nature, what we would now call scientific knowledge, could only be successfully deduced from the study of numerous individual examples, moving from the particular to the universal in a mode very similar to our modern notion of inductive reasoning: to extrapolate from the particular instance to the general law and only subsequently to work out how the general law is the explanation – the cause – of all the particular instances of it.

It was an approach that made empirical observation critical to the investigation of nature in fields like biology, because only by identifying and classifying relevant information about animals, such as what they had in common and in what respects they differed, is the philosopher able to discover what is invariably and necessarily true about living animals; and from this to deduce the causal relationship in play between a feature of the world and why it is that way: why do some animals and not others bear live young? What does life, the state of being alive, necessarily involve?

This recognition of empirical observation as the starting point of inquiry, so that definition and classification can provide true premises for the formal logical structures involved in Aristotle's notion of proof, means that this Aristotelian approach is one of the strands of Greco-Roman investigation into nature most similar to modern science. It included not only Aristotle's own work on biology, geology, physics and meteorology and his successor Theophrastus' inquiries into plants, but also the work of those subsequently influenced by Aristotelianism, such as the third-century BC medical anatomists Herophilus and Erasistratus.

In fact Aristotle discusses 'induction' in much less detail than he devotes to logical methods of demonstrative deduction, in which the validity is entailed by the structure of the argument, but the truth of which depends on its also having true premises. His more sketchily described process of moving from the particular to the general was a matter largely for those areas in which definitions and common notions could not be obtained without a systematic collection of information, on as large a scale as possible, about a particular phenomenon or category, as was generally the case in zoology or meteorology. In other subjects, such as mathematics or physics or logic itself, other modes of inquiry were more appropriate; and it is in respect to these that much of Aristotle's work on syllogistic logic, metaphysics and proof was carried out in works such as *Prior Analytics* and *Topics*. The 'inductive' approach was described in *Posterior Analytics*. Aristotle's biological researches can be found in his *Inquiry into Animals, Parts of Animals* and to some extent in *Generation of Animals*.

Therefore it can be said that Aristotle was the first person to prioritise and describe an explicit programme for a crucial component of scientific methodology: the systematic collection of information so that meaningful patterns can be sought within that body of data and explanations for those patterns thus derived. *The Oxford English Dictionary* defines scientific method as 'a method or procedure that has characterised natural science since the seventeenth century, consisting in systematic observation, measurement, and experiment, and the formulation, testing, and modification of hypotheses.' In the twenty-first century most of science still relies on the accumulation of particulars through observation and experiment: the development of statistics and probability theory has only maximised the need for large amounts of data.

Measurement or quantification is a feature of only a few aspects of Aristotle's scientific work; like most others in antiquity he used experiment even less. Certainty came from authoritative data, logically constructed argumentation and clear definitions, not from testing or Popperian falsification of hypotheses. Aristotle's work would not match the *OED* definition of scientific method. But he saw the importance of classification, in all subjects but particularly in biology. His taxonomic scheme divided animals into animals with and without blood (corresponding to vertebrates and invertebrates). Those 'with blood' were further divided into those that gave birth to live young or produced eggs, and those 'without blood' into insects, crustacea, and testacea. The concepts of 'genus' and 'species', developed from his work on logical relationships, also applied to biological relationships and reinforced Aristotle's perception of even the most chaotic parts of nature as ordered, hierarchical and fit-for-purpose, as well as his conviction that classification and investigation was the right method of discovery.

The concept of classification is important because it represents an acknowledgment, largely absent from the Pre-Socratics' speculations, that differences matter as much as similarities and that one explanation does not necessarily suit all species. The emphasis on logical forms of argument as the only sure kind of proof is also significant to the concept of science, because it establishes the idea that truth is only reliable when reached by well-defined, consistent routes that rule out alternative answers; and that even if scientific ideas are reached via the ladder of induction, they must ideally still be proven by other means. For Aristotle, that meant deductive logic, but in a less formal way it is important that statistical and scientific conclusions be what is called 'robust': that is, that they still hold even if certain variables or assumptions are altered.

Some of the specifics of Aristotle's zoological research are still accurate. He was the first person to observe that one of the eight arms of the male octopus is anatomically different from the others, because its function is to store and then transfer sperm to the female's eggs (*History of Animals* 4.1, 525a3–20). Aristotle got the purpose right as well, but his report was not confirmed or believed until the nineteenth century. His understanding of anatomy was often a dramatic improvement on that of his predecessors and medical contemporaries, because he carried out dissection upon all kinds of animals. (Not upon humans; consequently Aristotle's account of internal human anatomy had to be partly extrapolated from that of dogs or goats and was mistaken on those points where primate or human anatomy diverges.)

Aristotle's crucial insight into the use of dissection as a method of discovery was his realisation that structure is explicatory of function. Anatomical dissection proved to Aristotle that empirical investigation generates a better theoretical understanding, allowing one to see patterns and connect one aspect of nature to another. But it is sometimes a large step from making accurate observations of anatomy to making accurate deductions about the function of anatomical structures and Aristotle's efforts, like those of subsequent investigators, contained many misses as well as hits.

Dissection was necessarily supplemented by other sources of information about not only dead but living animals and plants. Much of this came not from Aristotle's own observations or those of men he had taught, but from people who had travelled abroad or those working in relevant fields, like bee-keepers and fishermen. Some of this information may have been told and retold several times, changing emphasis or losing precision in the process. Greece was still primarily an oral society, even among the small section of society that was literate.

Local knowledge of this kind is a sensible method of obtaining information before the information age and, as we saw in Chapter II, was still a feature of Victorian science, when Darwin consulted widely, if with some scepticism, amongst pigeon-keepers and far-flung ambassadors. And since Aristotle was more or less inventing the concept of studying animals and plants in this fashion, there was no already accumulated body of data to study.

But it also lacks an error-correcting mechanism beyond whatever the investigator finds doubtful. The consequence was that Aristotle drew conclusions about biological nature on the basis of a mass of information from sources of varying reliability, without paying much attention to any notion of corroboration and taking a lot of it on trust. He accepted 'common opinions' as reliable empirical data, since these were presumably based on the experience of large numbers of people.

One of his most famous errors was his assertion that females of any particular species have fewer teeth than males of that species. This includes humans, so according to Aristotle women have fewer teeth than men (*Parts of Animals* 2.3, 501b19–21). In fact, men and women have the same number of teeth. This apparently elementary mistake has puzzled many historians. Aristotle was married and his household must have contained female slaves, so nothing physically prevented counting teeth in examples of both sexes.

It is possible that whoever made this observation, whether Aristotle or someone else, based it on a very small sample, perhaps a single individual of each sex. Dental health in antiquity was much worse than it is now and many adults had less than their full complement of teeth, so that the number of teeth per individual would have varied much more than it would in a survey carried out today. It may even have been the case that women in antiquity actually did have on average fewer teeth than men. Robert Mayhew has argued for this on the basis of evidence suggesting that women of the period had a more restricted diet than men.[2] The belief that women had fewer teeth could then have been empirically founded in widespread experience. Alternatively, Aristotle might have come to this conclusion by a different process of reasoning. Aristotle consistently interprets biology in all species as demonstrating that the female is inferior to the male, so he could have viewed fewer female teeth as inherently plausible, just as females tended to be smaller and less physically strong.

Whichever one of these is right, the mystery of the women's teeth illustrates two things. Firstly, let us say that Mayhew is right and that women in Greek antiquity had lost more of their teeth than men. In this case Aristotle's data is empirically solid but his conclusion is still mistaken, because the data only shows that men on average lose fewer of their teeth then women do, not that they naturally have more teeth. In the second place, the mistake is less and less likely to be made if the sample size is increased. Even if women had lost more teeth on average, there would have been enough individual instances of men with fewer teeth for it to be obvious that the pattern was not a reliable one.

What all this demonstrates is that Aristotle was right about the usefulness of empirical investigation as the starting-point for theory: it provokes thought, constrains speculation and corroborates ideas. He had enough successes with this approach, particularly dissection, to be convinced of its explanatory power; and in fact empiricism is a robust enough approach that many of these successes were perfectly real, as real as the hectocolyd arm of an octopus.

But the empirical approach is a broad church. Science in our contemporary sense is a much narrower sect. Absent from Aristotle's study of nature are the techniques of modern science for avoiding and correcting error. Much of this involves procedures to reduce the numerous and varied ways in which human input can misinterpret information or bias its acquisition: hence the double-blinded, randomised trials of medicine, the emphasis on prediction before result, the notion of statistical significance and the quantification of anything possible. Modern science has learnt from its mistakes that more evidence is better, and different, corroborating, attack-proof, measurable, repeatable forms of evidence carried out by different people are best of all. As Robert Pursig put it, 'the real purpose of the scientific method is to make sure nature hasn't misled you into thinking you know something you actually don't know'.[3]

The approach of Aristotle, one of the most empirically minded thinkers of the ancient world, is by contrast positively laissez-faire, because Aristotle was not expecting nature to be misleading. The underlying assumption of empiricism is that that the world is largely how it appears to be and that its complications are susceptible to reasoned analysis on the basis of prima facie evidence. But a partial selection of such evidence can generate any number of reasonable stories about the world, all of which seem to be confirmed by observation and all of which can be wrong – for instance because much of the world is in fact invisible to the naked eye. Sometimes technology is the horse that has to come before the observational cart.

Moreover, it was not just that Aristotle expected the universe to be relatively straightforward. The patterns he saw assured him that nature as a whole was purposeful. The highest form of nature possible was the rational capacity of man, because this was the only intelligence capable of comprehending what had caused it to be. Aristotle's concept of nature was formed by the order he perceived in it, by his ability to derive consistent meaning in its patterns and even by the success of his empirical approach, but this all led him to the conclusion that the purpose of the universe was to be something perfectly understandable by its ultimate creation, the adult human male. He would not have expected that, in the words of the biologist J. B. S. Haldane, 'the Universe is not only queerer than we suppose, but queerer than we can suppose'.[4]

Of course, even modern science is predicated on the view that nature is something we can understand. We assume, disagreeing with Hamlet, that its mysteries can be solved by human reason; specifically, by scientific methods.[5] But much of what we know about the universe can only be expressed in mathematical terms. In some areas, such as string theory, these might be only conceptual possibilities. They are hard to test and lack empirical verification. If the hypothesis of the multiverse is correct, some physics could be true in other universes but not necessarily in ours (Chapter I). Even conventional physics and chemistry involve a series of metaphors for matter at the atomic level and sub-atomic level, because here the universe is entirely outside our direct experience. There is no way to describe something that behaves like both a wave and a particle, except to say that it behaves like a wave, but also like a particle.[6]

Humans evolved to have intelligence, but that intelligence was formed for specific social and environmental contexts. Its application to the mysteries of the universe has not always come naturally, hence the need for the error-correcting mechanisms of scientific method. But this would never happen in the Aristotelian viewpoint, because its universe is not only teleological, but anthropocentric.

Objects in Space

In modern science, a hugely important part of the process for avoiding error and corroborating speculation consists of experimentation. In antiquity, experiment is rare, even when it was technically feasible. A well-known missed opportunity concerned the speed at which heavier and lighter objects fall. Aristotle assumed that heavier ones fall faster than lighter ones and that weight was directly proportional to size, so that any object falls in a time inversely proportional to its weight.[7]

As a matter of fact, in atmosphere heavier objects sometimes do fall faster than lighter ones if there is sufficient air resistance to a lighter object with a large surface area, such as a piece of paper; or to an object moving very fast. The difference, however, is often too small to be observable to the naked eye, unless you are comparing a sheet of paper and a billiard ball, and the rate of difference is not proportional to weight. In vacuum, of course, objects of different mass fall at the same rate.

It is not unusual for ancient scientists (or modern ones) to have erroneous beliefs, but this is a particularly important example because it was within the bounds of ancient Greek technical capacity and imagination to test it. We know this because such experiments were actually carried out by the Christian intellectual John Philoponus (AD 490–570), author of an important but critical commentary on Aristotle's *Physics*. Aristotelian physics was the long-established paradigm, so Philoponus was working within much the same conceptual framework as his illustrious predecessor. It was also as technically feasible to carry out measurements of how long it took heavier and lighter bodies to fall a long distance in the fourth century BC as it was for Philoponus in the sixth century AD. The difference is that Philoponus bothered to carry out the experiment, and Aristotle did not:

> But this [view of Aristotle] is completely erroneous, and our view may be completely corroborated by actual observation more effectively than by any sort of verbal argument. For if you let fall from the same height two weights, one many times heavier than the other, you will see that the ratio of the times required for the motion does not depend [solely] on the weights, but that the difference in time is very small. And so, if the difference in the weights is not inconsiderable, that is, if one is, let us say, double the other, there will be no difference, or else an imperceptible difference, in time, though the difference in weight is by no means negligible, with one body weighing twice as much as the other.
> (Philoponus, *On Aristotle's Physics* 682–4)

Philoponus' experiment was part of a set of arguments against major points of Aristotelian physics. He was trying to overturn established theory. In this argumentative context, experiment represented extra ammunition. In contrast Aristotle was articulating a more general assumption that already fitted into his conception of physics and which he felt had no need of additional evidence. Ironically, given Aristotle's reputation as an advocate of empiricism in natural philosophy, he was still taking the universe on trust, whereas Philoponus had no trust in Aristotle.

But this refutation gained little traction in antiquity. Philoponus was unfashionable in other areas as well: after his death the Church excommunicated him as a heretic for his interpretation of the Christian

Trinity. His views survived because they were embedded in his commentary on Aristotle's *Physics* and such literary-philosophical analyses were highly valued in the intellectual world of later antiquity, to survive within a subsequent culture of scientific inquiry in the medieval Arab world. What did not count for much was the invocation of observation: no-one repeated or accepted what he saw by dropping weights under controlled conditions.

It was not until another millennium had passed that Galileo Galilei argued that bodies of different weights would fall at the same speed. Galileo, moreover, arrived at the idea that Aristotle must be wrong not through simply testing falling weights, but by demonstrating a logical contradiction in the Aristotelian theory. If a lighter and heavier object were tied together by string and the resulting combined artefact dropped from the top of a tower, the lighter object should fall more slowly and act as a brake on the heavier object, causing their combined weight to fall more slowly. Yet their combined weight is also heavier than the heavy object alone, so logically it should fall faster as well.

This is that pseudo-empiricist classic, the thought experiment. The more famous test of dropping a ten-pound and one-pound weight from the leaning tower of Pisa was probably also proposed as a test that would confirm Galileo's theory and disprove Aristotle's, rather than actually carried out.[8] The intellectual context at the time – give or take the Catholic Church – was more favourable to the overturning of scientific paradigms. Galileo's arguments spelt the beginning of the end for Aristotelian physics.

Objects in Motion (I)

Suppose someone takes a solid, weighty object and throws it upwards or sideways. The Greek word used by Aristotle for such an external motive force is *dunamis* or *dynamis*, which gives us the terms dynamism and dynamics.

The modern category of dynamics did not exist as such in antiquity, but related questions about movement and matter were articulated (not always very clearly) in Aristotle's textual works. We have already seen how Aristotle mistakenly assumed that heavier bodies fell proportionately faster than lighter ones, how in late antiquity Philoponus attacked this view on the basis of experimental observation; and how nonetheless Aristotelian physics remained the dominant paradigm of explanation until Galileo revived logical and experimental demonstrations of its falsity in the more receptive intellectual context of the sixteenth century.

This history is similar to that of another set of problems about moving objects, which was again answered somewhat differently by Aristotle, Philoponus, the Arab scientist-philosopher Avicenna or Ibn Sīnā (AD 980– 1037) and eventually Galileo and Newton, among others. This is a story in which the later protagonists are in dialogue with the opinions of their predecessors. It is possible to read it as a story of stepping stones to scientific success via successive paradigm shifts, in which Aristotle's contact causality provokes the 'impetus' theories of the late antique and medieval intellectuals and is finally replaced, through work of the sixteenth to eighteenth centuries, by the quantifiable concepts of inertia and momentum and their role in classical mechanics. Thomas Kuhn identified the medieval idea of 'impetus' as just such a paradigm breaker in his influential *The Structure of Scientific Revolutions* (Chicago, 1962). It is also possible to view the history of dynamics as a case in which the questions remained the same, but to which there was no satisfactory answer and much conceptual confusion until natural scientists like Galileo and Newton carried out experimentally orientated, quantified, repeated and detailed studies of how objects in motion actually behave, and were consequently able to reframe the subject under discussion in a useful way.

Consider again what happens when someone takes an object – say a javelin, the usual ancient example, or a ball or a stone – and throws it.

According to Aristotle, objects move according to the natural motion and natural place of their constitutive elements. For terrestrial objects, the four elements are earth, air, fire and water. Heavier objects contain more earth than lighter ones, which is why they fall faster (according to Aristotle), since earth naturally moves to the centre of the universe. Water is naturally situated just above earth, which is why springs bubble up and rain falls down. Very light objects are full of air, so they naturally maintain themselves in that region or fall slowly – like a leaf or a feather – if they contain a small amount of earth.

In Aristotelian physics, the speed of a moving object is proportional to the weight of the object and inversely proportional to the density of the medium: it is more difficult to swim through earth or honey than through water. An object only moves if it is not already in its natural place; or if it is acted upon by an external force that temporarily overwhelms the object's own natural motion and separates it from its natural place. All motion of this second kind is unnatural motion, because it is caused not by something intrinsic to the object-in-motion, but by an external force. If I pick up a stone, I remove it by force from its natural place and am acting against its natural motion. This division between the intrinsic and the extrinsic, the natural and the forced, is crucial to Aristotle's understanding of motion. He analyses all forms of movement in causal terms, categorising that cause as natural or external.

In *Physics* Book 8 Aristotle says that an external force moves an object, as when I pick up a stone, for as long as it is in contact with it. When the moving force ceases to be in contact with the stone, it ceases to cause the stone to move. But the stone's location in the cosmos has now shifted and it is no longer in its natural place. When the external force ceases to act directly on the object, the natural motion of the object is able to reassert itself. I let go of the stone, and it falls back to earth. If I tug on the string of a balloon, I pull that balloon, which consists mainly of the light element of air, downwards against its natural motion and away from its natural place. When I let go this time, the natural motion of the air causes its self-motion upwards again (taking the earth-heavy envelope of the balloon itself with it against its own natural motion). If the propulsive force does not cease, but ceases to be in contact with what it is moving, the object again stops moving. Unnatural motion lasts only as long as the initial force is applied and in contact with the object being moved: what one academic calls 'contact causality'.[9]

So far, so good, but what happens when I not only pick up a stone but throw it upwards and away from me as hard as I can? This looks problematic for Aristotle's system, because the unnatural motion of the stone continues even when I am no longer exerting my own force. What is making the stone move upwards when I am not forcing it upwards against its natural downward motion? And why does that unnatural motion cease a few moments later, without any other force apparently exerting itself, so that the stone does eventually fall back to earth?

Aristotle discusses this particular problem both in the *Physics* and again in his treatise *On the Heavens*. In both cases his treatment of the case of the thrown projectile is compressed and often obscure, causing difficulties for its interpretation in both later antiquity and modern scholarship. He says: [*kinda just ignores things when they don't fit his theory*]

> Further, as things are, things that are thrown are moved when they are not touching what pushed them, either because of circular replacement, as some say, or because the air pushes the-thing-that-was-pushed by a motion faster than that by which it is carried to its proper place.[10]

The first option here, 'circular replacement' (*antiperistasis*), a theory also found in Plato, sought to explain how motion occurred in a universe with no void or vacuum. Aristotle found the idea of nothing as a component of the universe to be logically and mathematically vacuous, and mustered numerous arguments against it. Instead of an object moving into empty space, in circular replacement the object and whatever was filling the space into which the object moved are displaced together, as partners exchange position in the steps of a dance. Its advocates appealed to observational confirmation. For instance, they pointed out, if you shake or stir a liquid, the individual material parts of that liquid move about but as a whole the volume and position of the liquid in the container does not change. Its constitutive elements, or molecules, have simply shifted places, but all places remained full and occupied throughout the process. As Aristotle elsewhere described it: 'For it is possible for things to yield place to one another at the same time, even though there is no distinct space in between apart from the moving bodies. And this is clear in the case of whirls of continuous things, just as it is in the case of those of liquids.'[11]

His phrasing in the *Physics* passage just quoted implies that the simultaneous displacement of *antiperistasis* is an alternative to a slightly different theory, in which the movement of the air in contact with the stone is faster than that of its natural motion and overrides it (or if the stone is thrown downwards, increases its natural rate of descent). These two explanations can be made compatible, which is perhaps why Aristotle seems unconcerned to explicate further or to say which he prefers. The basic idea would be that the original motive force affects not only the stone it is in contact with, but also the adjacent air. (Or the original contiguous air might be that in front of the projectile, displaced by it and triggering the next series of motive displacements. Philoponus objected to the way in which Aristotle's description seems to require air displaced in front of the arrow to move back along the side of the projectile, fill the space behind it and thence impart an extra motive force to the projectile.)

Air is naturally light, and so is moved faster and further than the heavy stone. The contiguous air is set into motion on its own account, hits the next segment of air and causes that to both move and be able to move the next segment of air; and so on in a continual process of displacement of air around the stone, carrying it with it or pushing from behind.

We can try to clarify Aristotle's thought here by looking at another passage on projectile motion from the *Physics*. It is always possible that Aristotle's thinking changed over the course of his life, so we cannot assume that he meant precisely the same in the *Physics* as in *On the Heavens*, but it does appear probable that the two texts are explicating much the same theory. *Physics* goes into more detail on the role of air, which Aristotle explains as an element both heavy and light, so that:

> Insofar as it [air] is light, it will make the locomotion upward whenever it is pushed and takes the source [of motion] from the

power [original mover: person throwing the stone] and again, insofar as it is heavy it will make downward motion. (8.10, 301b23–5)

The original external force making the stone move is my throw, but this ceases as I lose contact with the stone. The external force now operative on the stone is that of the air. Aristotle adds (267a2–5), 'we must say that the first mover makes, for example, air or water or some other such thing, which by nature moves and is moved, move [the stone]'. This point is very obscure, but, as we said, the original force of the throw also acted on the contiguous air, which became a set of contiguous air-segments successively moving on their own behalf, each of which in turn exerts a brief force on the stone, carrying it along with its own motion, in a sort of pass-the-parcel arrangement. The movement of the air becomes steadily less forceful, until in the end it moves the next segment of air but is not enough to make that segment something that causes motion. The stone is at this point no longer moved by an external force and its natural motion is no longer overwhelmed. The stone drops to earth.

Exactly what the original force does to the air is unclear. What Aristotle is not claiming is that the motive force of my throw was transmitted to the air and thence to the stone. The earliest notion of such a transmitted force, that is impetus, seems to be that of Philoponus himself and it was some time before this evolved into the inertia and momentum of classical mechanics, as discussed below, 'Objects in Motion (II)'.

One interpretation is that Aristotle is thinking of air or water or a similar medium as being in some way elastic. This makes it uniquely able to move something even when its own motion has ended (*Physics* 267a5–7). The decreasing capacity of each segment of air, as it moves away from the original force, to be a cause-of-motion is a product of reducing tension in this elastic capacity, until eventually the air is too slack to cause motion in whatever it touches, and the stone falls.[12] Projectile motion is therefore only a phenomenon of such elastic media, which have the ability to cause motion even when they themselves are still.

71

The Aristotelian commentator Simplicius (AD 490–560), a contemporary of Philoponus', wondered why we needed to drag air into the picture at all:

> But if we say that the man throwing the missile transfers to the air a steady motion, why don't we say that this motion is given to the missile without having recourse to the air and therefore without our being forced to assume that it [the air] is not only moved but also moving [i.e. a cause of motion]?[13]

In Aristotle's universe we cannot do as Simplicius suggests because causality works by contact. The motion of the stone through the air is being caused by something acting continuously and contiguously on it. Take away the cause of motion and you have no motion. Take away the external cause of motion, and only the internal cause is left – the stone drops. Media like air and water, meanwhile, create special circumstances because their nature involves an easily triggered intrinsic capacity to cause motion in adjacent objects, if this capacity is itself activated by an external force. If I try to force a stone through the ground, however, the soil and rocks there have no such elastic nature, which is why it is tough to throw a stone underground. In its way, this theory is an empirical one, because it makes use of the fact that objects move differently in different media, but because Aristotle is thinking exclusively in terms of what causes motion, he does not stop to examine closely how objects in motion actually behave.

Philoponus, as we have noted, was not impressed by the sequential causation of movement by air-segments, sweeping objects up with air's intrinsic lightness or accelerating their falls with its heaviness. He asked his own readers to imagine an army that never launched its ballistic missiles, such as javelins, but merely balanced them on a narrow parapet and then set in motion 10,000 bellows close behind them, stirring up the air. Would the missiles fly or merely overbalance over the parapet and crash straight downwards?

As an alternative, he asked a question similar to Simplicius and imagined a theory in which motive force could be transmitted from its origin (I throw the stone) into the stone itself, rather than into the air outside and around the stone. It is a radical departure from Aristotle in that the stone, the thing moved, now has an (acquired) intrinsic motion, a sort of battery power. This intrinsic motion is apart from, and can be opposed to, the stone's natural inclination to go to its rightful place in the universe.

This notion, many have argued, is the rough beginnings of a theory of impetus and thereby a step in the direction towards classical mechanics. Thomas Kuhn thus identified theories of impetus as one of his famous 'paradigm shifts' in science, but if so it was a slow-moving shift, bordering on the glacial. In Philoponus' version, the transmitted impetus is thought of as having a natural expiry stamp. The original force imparts only a certain amount of self-motility to the object. When that runs out, the object is out of impetus and falls: that is why a thrown object does not continue in motion forever. In the eleventh century, as both the Arabic tradition and the Latin West developed variations on the theme, the Persian philosopher Avicenna thought of impetus as a permanent force counteracted only by external opposing forces such as air resistance or friction. This edges close to the vital reformulation of dynamics that is Newton's First Law: an object either at rest or in motion continues in that state unless operated upon by an external force.

Eventually, the close-but-no-cigar theory of impetus gave way to the key concepts of inertia and momentum. These emerged slowly, partly out of impetus theories, some of which influenced the young Galileo in the late 1500s. In the seventeenth century Galileo's work and, subsequently, that of Isaac Newton, led to the downfall of Aristotelian physics in its entirety. A critical component in this revolution was the use of observation and experiment.

Objects in Motion (II)

In Aristotelian physics, as I just said at considerable length, the effect of a force only lasts as long as that force is itself maintained on the object. But

Galileo experimented with rolling balls down an inclined plane and watching as the ball continued not just downwards, but up a further inclined slope, until it was back almost at its original height. The ball moved closer to its original height if moving on a smoother surface, an observation which suggests the conclusion that a moving object naturally continues moving and that a force (in this case, friction) has to be exerted upon it before its motion decreases to rest. Newton's First Law of Motion states this as: the velocity of any object remains constant unless acted upon by an external force.

This quality of objects, the degree of their resistance to a change in their motion, is inertia; and it is dependent upon the mass (or in pre-gravity science, the weight) of an object. The more mass an object has, the harder it is to effect a change in its motion, whether that is a change from rest to motion, from motion to rest, a change in velocity – slowed down or speeded up – or a change in its direction (more massive objects do not go round corners as easily as lighter ones).

The closely related property of momentum is a measurement of mass as it moves in a particular direction (an object at rest has no momentum). The amount of momentum that a moving object has is mass multiplied by velocity. It takes a lot of force, or a smaller force applied for longer, to stop either a small object of low mass moving fast or a massive object rolling slowly. A fast-moving massive object is a serious problem. When stopping a moving object, one applies a force against that object's direction of travel, but you can also change an object's momentum by applying a force in the same direction. That force still works against the object's inertia, its resistance to a change in state, but it increases the object's velocity and, consequently, its momentum.

Post-Aristotelian thought successfully abstracted away from the actual substance of the objects in question themselves. The chemical make-up of a javelin, a ball or an air balloon became irrelevant to a generalised understanding of the dynamics of moving objects. This was not because chemistry had improved, from thinking in terms of elemental constituents of earth and water to a concept of atomic weight (discussed below), but because experimental observation by Philoponus, Galileo and others had disentangled the nature of motion from the physical objects that embody it on a temporary basis, as well as from properties such as air resistance and friction.

Inertia explains why an object continues in motion even when the force of the original throw has ceased to apply, regardless of whether any object is earthy or airy, with their consequent natural place and natural motion. Momentum, which considers direction as a property of the object, explains how a moving object interacts with its environment, instead of thinking of particular elements as having a naturally fixed position in the universe. The only relevant properties of the object are its mass, speed and direction, which can be represented by the abstract terms of mathematics.

This also helps us to think about what is happening in terms of more than one object at a time. Take momentum again. In Philoponus' theory of impetus, the force transmitted by my throw or the arrow string to the projectile gradually ran down, like a battery. But momentum is what physicists call 'conserved' – its amount remains constant over the system as a whole. If two rolling balls collide on the pool table, one may lose some momentum, but the total amount within a closed system remains the same as before the collision.

In this process, even relatively simple experimentation with dropped objects, inclined planes and the careful variation of weights, distances and surface friction, can reduce the number of questions relevant to the phenomenon under consideration and disprove confusing assumptions, such as the faster velocity of heavier weights. Aristotle sought to explain dynamics as part of an entire theory of physics, religion, ethics and causation. He saw objects in motion as belonging to already established categories of cause: natural and unnatural. His theory acquired an inertial deadweight of its own through the succeeding centuries, but its obscure and unsatisfactory treatment of moving bodies attracted relatively early criticism and interest and created a target area for a more successfully focused approach.

Motion into Empty Spaces

Dynamics were not the only area of Aristotle's theory of motion to attract controversy and disagreement. Aristotle argued vigorously against the existence of void or vacuum: an extent of three-dimensional space containing nothing. He preferred the idea of motion as a simultaneous rearrangement of matter. This is a theory with mathematical consequences, since it requires matter to be infinitely divisible. Observation and experimental evidence played a role in the ancient debate on these matters as well, but it was not until the eighteenth century that experiments produced new evidence for atomism.

The usefulness of void as a concept is that it provided a readily believable explanation as to how motion is possible. If there was no empty space anywhere in the universe, not a millionth of a millionth of a millimetre, then there seemed to be literally no room for motion to even begin. *Antiperistasis*, the alternative theory of motion without void discussed above, is more counter-intuitive at this beginning point. The earliest atomists, Leucippus and Democritus, postulated empty space, that is void, as one of the two things that exist forever, with the other being the atoms that fill some of that space. We can imagine the atomist universe as a spatially infinite extent of emptiness through which tiny particles of solid matter move. The temporary collections of atoms that make up macroscopic worlds are evanescent pockets of greater and lesser material density within this emptiness.

Some subsequent thinkers flipped this notion of a space partially filled with particles upside-down, into a concept of matter dotted with tiny bits of emptiness. According to these theories matter above the elementary scale is full of tiny discontinuous pockets of empty space, an idea now called 'microvoid', in the same way as a sponge is full of little air pockets. This concept was probably developed by Strato of Lampsacus (335–269 BC), aka 'the physicist', who became the third head of the Lyceum school of philosophy after Aristotle and Theophrastus. Since Strato's own treatises are lost, we have to reconstruct the beginnings of the idea from what is said about both microvoid and Strato in later authors, particularly the mathematician-engineer Hero of Alexandria (AD 10–70) and the Aristotelian commentators from late antiquity. Simplicius, for example, thought that Strato's theory involved the following:

> Strato of Lampsacus tried to prove that void (literally 'what is empty') is interspersed in all bodies, so as not to be continuous, saying thus: neither water nor air nor any other body could be penetrated by light or heat or any other corporeal power. For how could the sun's rays penetrate to the bottom of the jar? If the liquid had no gaps in it and the rays entered with force, then the jars would overflow...[14]

In other words, if macroscopic matter were entirely solid, nothing could penetrate it. Since Strato understood light and heat as consisting themselves of particulate bodies, they would simply bounce off such matter, just as it is not possible to shine a light through a solid piece of wood. If air were similarly completely solid, it would be opaque instead of transparent and the same applies to water or any other transparent or semi-transparent medium. Matter, then, must contain space through which other kinds of material can move, as light passes through air. Strato conceived of matter as particulate – not in the atomic sense in which a macroscopic structure is the combined effect of interlocking solid shapes which may look nothing like the finished product, but in a sense in which matter would be like itself all the way down, so that wood or light or water consists of tiny particles of wood or light or water, fitted closely but not perfectly together. Hero draws a visual analogy with grains of sand on the beach, in which he says that the sand grains are analogous to air particles, and the air between the grains of sand to the empty spaces between the particles of air. In the interstitial joints of matter are these tiny gaps of emptiness, available to be filled by the particles of a beam of light or of air.

This theory could also provide an explanation for how sound travels through air and other media like water, and even through objects insufficiently porous for light. One can hear through a wall, for instance, because sound travels through a ripple effect of ongoing compression of air particles, moving the sound from one side of the wall to the other. Void is again necessary for the air particles to be able to get through the apparently solid matter. Wood can be burnt because particles of fire get inside it through such gaps, but stone is denser and the fire particles are too large and coarse, keeping heat and fire on the outside.

Hero pointed to other observational evidence for microvoid, or rather to familiar phenomena which microvoid plausibly explained. Add wine to your drinking-cup of water and the darker wine disperses through the water, suggesting that the water is not continuous – it can be broken up as wine moves into its interstitial spaces. Air can be compressed into a smaller volume than usual by forcing its particles closer together, into the gaps, but will seek to reestablish its normal structural arrangement, producing the effects of high pressure.

Hero described, and might have carried out, basic experiments on the same theme:

> If, then, we pour water into an apparently empty vessel, air will leave the vessel proportional in quantity to the water which enters it. This may be seen from the following experiment. Let the vessel which seems to be empty be inverted, and, being carefully kept upright, pressed down into water; the water will not enter it even though it be entirely immersed: so that it is manifest that the air, being matter, and having itself filled all the space in the vessel, does not allow the water to enter. Now, if we bore the bottom of the vessel, the water will enter through the mouth, but the air will escape through the hole. Again, if, before perforating the bottom, we raise the vessel vertically, and turn it up, we shall find the inner surface of the vessel entirely free from moisture, exactly as it was before immersion. Hence it must be assumed that the air is matter. The air when set in motion becomes wind (for wind is nothing else but air in motion), and if, when the bottom of the vessel has been pierced and the water

is entering, we place the hand over the hole, we shall feel the wind escaping from the vessel; and this is nothing else but the air which is being driven out by the water. It is not then to be supposed that there exists in nature a distinct and continuous emptiness [vacuum], but that it is distributed in small measures through air and liquid and all other bodies. (Hero, *Pneumatics*, Introduction)

In Strato and mathematician-engineers like Hero, the existence of vacuum was successfully decoupled from atomism. It still involved, however, a particulate, discontinuous theory of matter that contrasted with the infinitely divisible material of the Aristotelian universe. It was not until the eighteenth century that experiment readjusted the terms of the debate, and did so in favour of the particle alternative, in the shape of the British chemist John Dalton's revived theory of atoms.

In Hero's treatise, the existence of void is demonstrated through a real-world illustration. This is an experiment in the 'try this at home' format associated with science education on television. It is less an experiment in the sense of novel discovery or testing a speculative hypothesis and more an attempt at convincing an audience through appeal to something that the author is confident is the case. In the example quoted from *Pneumatics* Hero does not go far beyond the kind of knowledge familiar from everyday experience. A perforated container is all that is required in the way of special equipment.

That is not to say that more systematic experimentation, aimed at improving, testing or discovering new ideas and information, did not exist in Greco-Roman times, but it was relatively rare. One example is in ballistics development. Catapult-makers tried to work out the design of an effective catapult and how to construct this in larger and smaller versions without losing efficiency, according to an approximating formula based on spring diameter. This formula – the ideal ratio between the diameter of the hole for the catapult spring and the size of the catapult – was established by trial and error: a form of open-ended experimentation.[15]

The evidence that atoms were the right way to think about matter below the level of the macroscopic came from a procedure available in antiquity, but with an improved technical apparatus that greatly increased the sensitivity and certainty of its results. Equally important was the emphasis on experimental discovery of the seventeenth and eighteenth centuries, which owed something to the preceding centuries of alchemical investigation.

The technique involved using the fact that different substances have different physical-chemical properties, such as mass and density, which are consistent for each example of a substance. Gold is always heavier than the same amount of lead. The well-known story of Archimedes proving that the golden crown of the Sicilian king was actually an amalgamation of gold over less dense silver relies on this fact and on Archimedes' principle about floating and submerged bodies. The volume of the crown was difficult to establish – it was presumably one of the more jaggedly irregular kind of crowns – and could not be melted down, so direct comparison with unimpeachably solid gold of the same amount was impossible. By Archimedes' principle, however, the volume of water displaced by the submerged object will be the same as that object's own volume, and the object's mass can then be established by dividing the crown's weight (its mass under gravity) by its volume, to establish its density. That could be compared with the density of gold. The crown's density did not match.[16]

By the eighteenth century there was a more precise mechanism for weighing the weight or mass of an object, even when that object was a small amount of an element or chemical compound.[17] The chemical or analytical balance, now sometimes also called the atomic balance, is accurate to about one in 10,000. It enabled the French chemist Antoine Lavoisier to establish a quantitative basis for chemistry, by means of his careful measurements of numerous substances, elemental and compounded, both before and after chemical reactions.[18]

Lavoisier's experiments repeatedly demonstrated that the sum total of the weights of the reactants, the original materials in a reaction, was the same as the total of the weights of their products. These experiments are in principle simple, but in practice required time, experience with chemical reactions and attention to detail. In one example, Lavoisier formed a red mercury oxide by heating mercury in a sealed flask that also contained air, a process that took several days. The remaining gaseous air (nitrogen) in the flask weighed less than it had at the start (subtracting the weight of the flask) and lacked properties of normal air: animals could not breathe it and flames went out.[19] In the second stage of the experiment, Lavoisier took some of the mercury oxide, weighed it, and heated it again, producing oxygen and reducing the mass of the oxide. The weight lost by the oxide was equal to the weight of oxygen it had produced: the total weight of the original mercury oxide was unchanged in the sum of its byproducts after the reaction had taken place. In a set of related tests he was able to show that when oxygen reacted with other elements they gained weight in the same quantity as oxygen lost it.

Lavoisier had discovered a natural law, that mass is conserved in a chemical reaction. 25 years later, in 1803, the English scientist John Dalton realised that this and other laws established by chemical experiment could be most easily explained by a theory of discontinuous, atomised matter. He suggested that each basic element consisted of indestructible atoms whose properties, including their weight, cannot be changed.[20] This accounted for the conservation of mass through a chemical reaction. Take, for instance, the conversion of one molecule of methane and two of oxygen into one molecule of carbon dioxide and two of water. At the start of this reaction is one atom of carbon, four of hydrogen and four of oxygen. At the end of this reaction is one atom of carbon, four of hydrogen and four of oxygen. The atoms' total and individual mass and weight is also unaltered. The chemical change is a rearrangement of the bonds between these atoms, not a change in the atoms themselves.

Conservation of mass was not the only observable law for which Dalton's new atomic theory could successfully account. Experimental chemistry of the period, by measuring the mass of many compounds, provided evidentiary support for a law of nature postulated by Joseph-Louis Proust (AD 1754–1826), called the law of definite proportions or law of constant composition. This says that the proportionate mass of the constituent elements in any compound is stable and always the same, regardless of how or where that compound is formed.[21] At the time it was a controversial alternative to the idea that elements could combine in any proportion to form a distinct compound. Proust's law clarified the emerging distinction between a mixture, in which elements can be in any proportion to each other, and a compound, i.e. a new substance produced by a chemical reaction, in which the proportion of the elements, measured as their weight or mass, is always the same. The chloride in sodium chloride, for example, is always 60.66 per cent or about 3/5 of the mass of sodium chloride; oxygen is always 8/9 of the mass of any amount of water and hydrogen the remaining 1/9.

Dalton's own work, moreover, led him to the law of multiple proportions, which relates to those cases in which two elements can form more than one compound. For example, oxygen can combine with carbon to form both carbon dioxide and carbon monoxide. But for a fixed mass of one of the elements involved, say 100 grams of carbon, there can only be a reaction with either 133 or 266 grams of oxygen. Carbon dioxide is produced by 133 grams of oxygen and 100 grams of carbon; but 266 grams of oxygen and 100 grams of carbon results in carbon monoxide. The ratio of the reactant amounts of oxygen is 266:133, or 2:1. This pattern holds for other such compounds, enabling one to say that for a fixed amount of one element, the reactant amount of the other element forms a ratio of small integers, such as 2:1.[22] These two laws are the basis of stoichiometry, the quantitative relationships between participants in chemical compounds and reactions.

Dalton's insight was to see that these fixed proportions, coupled with the conservation of mass, made sense if you thought of an element's matter as consisting of a whole number of atoms of the same weight, which exchanged one or more of their own number for an equivalent weight of atoms of a different substance.[23] The fact that atomic ratios are not fractional leads to the conclusion that Dalton drew, that matter is not continuous and infinitely divisible, but consists of countable discrete items that cannot be further divided. Methane is the product of combining carbon with hydrogen in the ratio of 1:4 (CH_4), but nothing on earth or off it has the carbon to hydrogen ratio 1:1.4123. The units of matter that persist through chemical reactions come only in whole numbers.

Dalton took the name for his indestructible, indivisible units of matter from the ancient Greek theory. His atoms share those characteristics with those of Democritus and the Epicureans. Both are also examples of discontinuous theories of matter, along with Strato and the engineers' idea of microscopic elemental particles.

In a way, Dalton's theory set out to do less than ancient atomic theories, or was at least more narrowly focused. His original research concerned gaseous transformations and eventually chemical reactions that formed compounds. In particular, it accounted for a particular set of observations about conservation of mass and mass ratios in compounds.

In contrast, Democritus and the Epicureans used atomism as a one-size-fits-all explanation for every physical, social, perceptual, religious and ethical question, using mainly arguments derived from logic, analogy and criticism of alternative theories. The claim that a reductionist, material conception of the cosmos successfully explains such dilemmas is a reasonable position in a general sense, but applying atomism specifically and at every level, from the areas of study now filled by chemistry and physics to those of physiology, neurology, and meteorology is an exercise in ingenuity with little real-world corroboration. Consequently, its appeal depended largely on how its audience felt about reductionist, material and non-religious ideas in the first place.

Dalton's theory was equally wide-ranging and important in its implications: few subjects are more general than the nature of matter. It represented a triumph for reductionism and for materialism, however, because it was extremely convincing in its original, relatively narrow, field of inquiry: how materials actually behave. Dalton's conclusions have been most successful in this relatively circumscribed arena, under what might be called everyday earth conditions. Under more extreme conditions, such as those in high-particle accelerators, atoms are not indestructible. Isotopes mean that elements can have different relative atomic weights (because the number of neutrons vary). Some complex compounds mean that the integers of the law of definite proportions are not all that small. But these have extended rather than contradicted atomism as a theory of physical chemistry as it usually happens. Antoine Lavoisier defined an element pragmatically, as whatever could not be further broken down, and this nod to the contingent nature of scientific discovery and the complexity of the world is missing from the Greek search for the perfect and unchanging answer.

The quantitative basis of physical chemistry that these scientists established through experimental observation redefined the relevant questions in precise, measurable terms and excluded some of the answers, such as whether mass can be lost from a system or whether a compound's elemental proportions are variable. Within these parameters, considering only the mass and volume of particular elements, in controlled experiments separated from the confounding factors of the real world and by doing such experiments in sufficient number to establish patterns of behaviour, the chemists of the seventeenth and eighteenth centuries were able to find enough evidence to demonstrate the atomic nature of matter. Their focus on the particular enabled an understanding of something much more general.

Paradigm Shifts

The picture of the universe that resulted was a surprisingly empty one. Strato's notion of microvoid had to be more or less inverted. In that story, most of the universe was full of matter and empty space existed only in the tiny fragmented spaces between atomic particles. Stratonian–Heronian physics rejected the atomists' idea of an infinite expanse of cosmic void. The atomists themselves thought of void as the dominant constituent only in this larger cosmic sense. Within the confines of a world, whether this one or another, void was a more interstitial affair that gave atoms literal room to move but, conceptually, formed the subordinate backdrop against which the drama of material movement, collision and arrangement took place.

In the modern cosmos, it is matter that is hard to find in the expanse of spatial–temporal nothingness. It ghosts into uncertain, scattered and ephemeral existence as if someone has thrown a handful of grains of sand into the vastness of space.[24] Stars, planets and dust clouds are isolated conglomerations of matter that are at their closest millions of miles apart. The sun of our solar system is about 93 million miles from the Earth, and 3.67 billion miles from Pluto. The nearest star, Proxima Centauri, is 4.2 light years out, when a light year measures nearly 6 trillion miles. The nearest galaxy, the Canis Major Dwarf Galaxy, is about 25,000 light years from this solar system. It is a part of what astronomers call the Local Group.

This is a picture that Democritus could have recognised, although perhaps his cosmos might have been more misty with clouds of colliding atoms. It is not only on the cosmic scale, however, that the universe is startlingly empty. The twentieth century discovered that the atom itself was largely a matter of void. The space around the atom's nucleus is about 10,000 times the diameter of the nucleus, and it is empty apart from a single-digit number of electrons. Since this space is defined as the area in which there is a probability of finding an electron at a particular point, and this only reaches zero at an infinite distance from the nucleus, there is a sense in which atoms are infinitely large.[25] Most of the matter of an atom is concentrated in the relatively tight-packed space of its nucleus, where the neutrons and protons jostle each other and make up 99.9 per cent of the atom's mass.

Of course, at the atomic level the actual distances involved are to us tiny; much smaller than the space between grains of sand on a beach that Hero invoked as analogical to the emptiness of microvoid between jostling elemental particles. But inherent in that image is the idea that particles of matter are close to each other, so that the distance between them, the extent of the void, is thought to be less than the distance across the solid, unbreakable, part-less atom. Contemporary atomism, built originally on Dalton's insight but itself transformed by more recent discovery, reveals a world much more at odds with how it looks to us, and much less like grains of sand on a beach. From the atom to the cell, the animal and the planet, as well as the universe itself, void is almost everything. Humans consist largely of nothing. On every scale matter is strung out across great expanses of non-matter, held loosely together by energetic bonds of variable and breakable strength.

This counter-intuitive shift in perspective from the ancient is accompanied by another and more famous theoretical change. It turned out that not even atoms are forever. Splitting the atom opened up a new and different world for physics and, by accessing the power of the forces of the atomic nucleus, a new and different world for everyone.

This does not itself invalidate the ancient idea, developed as a logical answer to the problem of explaining material continuity simultaneously with material macroscopic change, of an unchanging unit of matter that combines temporarily with others. Arguably it simply pushes the idea one step further, as if matter was a series of matryoshka Russian dolls, each one containing smaller particles of matter. So far this merely confirms the usefulness of Lavoisier's contingent definition of an element as whatever cannot (yet) be further broken down.

But there is a limit to how often we can break down the latest candidate for true atomism in the search for fundamental elements of matter without finding the whole conception seriously inadequate. Discovering that Dalton's candidate for 'atom' has particles and parts of its own is one thing, but even an electron does not really behave within the atom simply as a solid particle of matter moving around in empty space. The familiar school textbook picture of orbits and nucleus is so far an over-simplification as to be actively misleading. And in the sub-atomic world of particle and quantum physics, we are not so much in a matryoska series as down the rabbit hole and heading fast for a mathematical wonderland. Lewis Carroll might have originally written *Alice* as a satire on non-classical mathematics of his own day, but the story of a world with an internal logic quite at variance with normal concepts of size, causality, change, place and time is not a bad image for where we are now.

Our imagination and to a large extent our logic are both derived from experience of how things work at the macroscopic level. But two millennia or so of searching for laws of nature has revealed that 'nature' is extremely diverse, and that what works in studying one segment or level of it may not work for others. This is not only so in terms of different research methodology in, say, biology and physics. It has also turned out that some theories only work within certain limits: Newtonian classical mechanics can get you to the moon, but as an explanation of the cosmos it has been superseded by Einsteinian physics.

Aristotle grasped something of this point, but not how deep the diversity might run; and his anthropocentric view of nature underestimated its alienness to human perspective. Science abstracts and simplifies the natural world in its attempt to understand and predict it, but scientists have been, and are, often tripped up by their personal and cultural assumptions, preferences, metaphysics and limitations. Science's historical journey to explanatory and technological success has been a piecemeal one in which detailed and diverse investigation of small areas, using experiment, statistics, repeatability and technology, has been productive as a means to understand one thing at a time. In antiquity, the argument of natural philosophy was that the universe could be understood through investigation and logic. They underestimated just how much investigation and what unusual forms of logic would be required.

CHAPTER IV

ILLNESS AND DISEASE

Medical Cultures

Different cultures often understand health, illness, disease and medicine in different ways. Several such medical cultures, with different systems of explanation and treatment, can co-exist within one nation or society. The result is that several models of medicine – that is, how to prevent and cure illness – compete, both as a means of attracting patients and as systems of explanation. In modern Western society, for example, there is a dominant medical paradigm practised by state-licensed practitioners who have gone through a particular training process and qualification. The medicine they practise, sometimes labelled 'biomedicine', is based upon the methods, evidentiary criteria and theories of modern science. It is widely accepted and socially authoritative.

At the same time various forms of 'alternative medicine' are also available, which offer explanations and treatments either unproven by scientific methods or actively in conflict with scientific theory. These are sometimes relatively new, like chiropractics (from the end of the nineteenth century) or therapies citing electromagnetic fields. Many others derive from ideas that originated and became well-established before the modern scientific era, but in which people still believe. Several are strongly associated with a particular culture. 'Chinese' medicine, including herbal remedies derived from China and acupuncture, is an obvious example and in the West attracts many non-Chinese clients.

We can also draw a distinction, even within one system of medicine, between the expectations and understanding of the patient or potential patient, and those of practitioners. Believing (or at least not disbelieving) in the efficacy of a treatment does not entail knowledge of how and why that treatment is thought to work by the practitioners, or by what methods diagnosis and decisions are made. Anthropologists use the terms 'illness' and 'disease' to draw a distinction between how people experience illness and what its causes and nature are understood to be. The latter category is the concept of 'disease' employed by medical practitioners in biomedicine or any other medical culture.

The practitioner's theory might also be understood and accepted by their patients, but often is not, at least not in any detail. Most people take antibiotics less because they understand how they work than because an accepted authority has offered them, and because past experience and anecdotal evidence suggests that people get better after taking antibiotics. Taking pills is a cultural norm and the most common outcome of any interaction with professional medicine. This is why patients often demand and get antibiotics for illness caused by a virus, in spite of the fact that they will not work. The experience of respiratory illness does not distinguish between the causative factors involved.[1]

This phenomenon, of different medical systems existing within the same society at the same time, can be observed in antiquity. We tend to think of Greco-Roman medicine, as practised by Greek physicians from the fifth century BC onwards and expanding with Greek colonists, conquerors and craftsmen into the east, the west and eventually the world of the Roman empire, as being a forerunner of current scientific medicine. Influenced by Greek philosophy, it utilised the concept of 'nature' found in contemporary philosophy and rejected the role of the gods in disease causation in favour of a model of illness based upon normally occurring material substances and their effects on the structures and mechanisms of the human body.

But an account of medical practice in antiquity that focuses on this 'naturalistic' or 'scientific' medicine, as histories of science usually do, excludes other forms. The literally home-grown medicine of household and family, consisting of remedies passed down through local oral tradition and taking most of its material ingredients from the foods, plants and animals accessible from ordinary diet, hedgerows and farming, frequently involved the use of ritual words or gestures. Just such a folk medicine of food-based medicines and 'magical' incantations was embedded within Roman aristocratic culture as strongly as it was amongst Italian peasants. The elder Cato (234–149 BC) recorded many such medical remedies. Dislocation or fracture, for example, was treated by a ritual that involved the splitting and recombination of a green reed applied to the injured limb to the accompaniment of nonsense words: 'Begin to chant: "motas uaeta daries dardares astataries dissunapiter" and continue until they meet.' Of cabbage, he says: 'It promotes digestion marvellously and is an excellent laxative, and the urine [of a cabbage-eater] is wholesome for everything.'[2]

Specialist 'purifiers' and 'diviners' competed with physicians for patients, while what we now call 'temple medicine' was a familiar and socially accepted option. This refers not just to invoking the aid of a god or gods at a particular temple, but in particular to the practice of 'incubation' at the sanctuaries of the divinised medical hero Asclepius. In incubation, pilgrims seeking help from the god made an offering to him and slept overnight in a special area of the sanctuary, hoping that the god would visit them in their dreams to grant a cure or give them medical instructions. Some of these dreams were memorialised in narrative form at a subsequent date and displayed as inscriptions on stone monuments in the sanctuary grounds. They helped to create people's expectations about what dreams of the god were like and reinforced the idea that cure followed dream. One example reads:

> Arata, a Spartan, suffering from dropsy. On her behalf her mother slept in the sanctuary while she stayed in Sparta. It seemed to her that the god cut off her daughter's head and hung her body with the neck downwards. After a considerable amount of water had flowed out, he released the body and put the head back on her neck. After she saw this dream she returned to Sparta and found that her daughter had recovered and had seen the same dream.[3]

Many patients, including those from the social elite, sought curative dreams from Asclepius as well as treatment from their physician. The practice spread from southern Greece to all over the Greco-Roman world. The Roman orator and writer Aelius Aristides (AD 117–89) managed his chronic experience of illness over decades through the often cryptic dreams sent by Asclepius, as interpreted by Aristides, his doctors, the officials of the cult and his friends and family. Such folk, religious or 'magical' systems of explanation and healing enjoyed longevity and popularity within Greco-Roman society and often also had a significant degree of intellectual authority.

It is thus easy, by identifying naturalistic medicine as a precursor of modern biomedicine, to divorce our understanding of Greco-Roman medical texts from their contemporary context. Yet naturalistic medicine is not just an origin story for a more successful modern counterpart. It is revealing, in its own right, of the beliefs, concerns and epistemic approaches of the Greco-Romans themselves.

Naturalistic Medicine: An Influential Meme

A fourth-century BC text on medicine, known as *On the Nature of Man*, articulated the theory that the human constitution consists of four fluid substances: blood, phlegm, yellow bile and the mysterious black bile.[4] In a healthy person, these substances and their particular qualities of hotness, dryness, wetness and coldness, are in balance. Illness is the result of imbalance. Medicine restores humoural ratios to a healthy or normative state, or it prevents an imbalance from occurring. Inherited or environmentally-caused variations in an individual's humoural balance explain body type, personality and predisposition to illness or congenital disease.

For some time, this model was only one among many theories of human physiology. Humours were a common explanatory concept, with links to philosophical ideas about the cosmos' fundamental elements and their qualities, but their number and sub-types varied. On the other hand the treatise *Breaths*, again from the end of the fifth century BC, attributed all disease to the ingestion or incorporation of air; and later Erasistratus of Ceos (*c*.304–260 BC) offered a mechanised model of the human body and its processes in which the movement of blood and air through tubes and valves, and blockages caused by the inefficient digestion of food into nutritive blood, were the crucial elements. Much later, however, the Greco-Roman physician Galen of Pergamum (AD *c*.130–216) adopted the *On the Nature of Man*'s four-humour theory for his own. This was in line with his systematic presentation of works (including *On the Nature of Man*) that were attributed to the fifth-century BC physician Hippocrates of Cos, as authoritative medical truth – once the ideas of 'Hippocrates' had been interpreted, updated and improved by Galen himself (this is discussed further in Chapter VI). Galen's personal success in medicine and prodigious textual output eventually resulted in his works dominating the transmission of medical knowledge in late antiquity. Galenism remained central to medical orthodoxy in the West through the medieval period and, in many respects, until the mid-nineteenth century. The four-humour theory thus became integral to this increasingly standard model of human physiology.

Another text ascribed to Hippocrates was the work now known as *On the Sacred Disease*. In antiquity, illness involving dramatic seizures (presumably epilepsy and similar conditions) was often thought to be caused by the gods, so the individual thus afflicted was perceived as polluted in some way. Cures aimed at the purification – the Greek word is *catharsis* – of sufferers through a variety of means. These included avoiding certain substances in one's diet or certain kinds of clothing, as well as ritual and religious purification.

The author of *On the Sacred Disease* argued polemically against this understanding of the condition's cause, and consequently against the healers and ritual experts who accepted the gods' role in it. Instead, he ascribed the illness to an excess in the brain of the humour phlegm: by chilling the blood and blocking the passage of breath or air through the blood vessels it interfered with the senses, the voice and the mind's control over the body.

ILLNESS AND DISEASE

This text has been central to the modern understanding of the kind of medicine and natural philosophy articulated in the Hippocratic texts and subsequently integral to much of Greco-Roman medicine. To a large degree this is because it explicitly rejects the role of the gods in disease causation. It replaces this model of illness with an explanation based upon normally occurring material substances and their effects on the physical operation of the human body.

This formulation, *mutatis mutandis*, is a familiar one in the modern world: whatever one's stance on whether god(s) exist, they have no causal role within the scientific understanding of physical processes. Miracles either supervene on those processes (if you believe in divine intervention) or do not occur at all (if you do not). Data and experience can instead be adequately understood, and cures made, on the basis of a causal model in which invariable physical laws act on and through materials. *On the Sacred Disease* offers a similar approach. Its author took it as read that the gods existed, but argued that it was a mistake in theology to think they could cause disease: because disease is pollution embodied and gods are pollution's antithesis. Thus the gods are essentially irrelevant to understanding disease, which is instead explicable in terms of a describable and predictable physics of the body.

As we saw in Chapter III, our term 'physics' is derived from the Greek word *physis*, usually translated as 'nature', as in natural philosophy or natural history. One might think of it as what normally occurs without outside intervention. Its use in antiquity among philosophers, medical thinkers, and other intellectuals also tends to contain two assumptions. Firstly, that what 'happens naturally', or 'nature', is consistent: the same effects always follow the same causes. Secondly, that human observation and reason is capable of identifying, understanding and predicting what is natural and, therefore, of manipulating it – for instance, altering the balance of one's humours through diet. It is therefore strongly rationalistic in its approach; that is, it assumes that reason is a useful method of proceeding that arrives at consistent and universalisable truths. It tends to exclude things which do not fit easily or necessarily within this paradigm, such as gods with personalities, or invisible mechanisms of disease like pollution/contagion (for the last see 'The Contagion Superstition' later in this chapter).

The 'Hippocratic' texts are united by this 'naturalistic' point of view, either explicitly or implicitly. But the distinction between naturalistic explanations and religious ones is not quite as clear-cut as I have made it sound. The primary objection of the author of *On the Sacred Disease* to gods as causes of disease is, as we have just seen, a theological one. Purifiers have got gods wrong. He is also willing, in a rather non-committal manner, to acknowledge that gods may help to cure illness, as long as they are consulted through the standard, acceptable means of prayer and visits to their temples. The gods are not rejected wholesale but their importance is diminished and the sphere of their activity is confined and constrained to traditional, socially validated forms of religious activity, with which Greco-Roman doctors were not willing to explicitly compete or to oppose. *On the Sacred Disease* instead directed its fire at its direct competitors in medical practice, in particular the purifiers. In doing so, it implicitly suggests that such healers were commonplace and their views on divine causation widely believed.

Similarly, one of the most effective arguments that *On the Sacred Disease* makes against purifiers is the accusation that their diagnostic classifications are arbitrary and unsupported:

> They make a different god responsible for each of the different forms of the complaint. If the sufferer acts like a goat, if he roars, or has convulsions affecting the right side, they say the mother of the gods is responsible. If he utters a higher-pitched [...] cry, they say he is like a horse and blame Poseidon. If the sufferer should be incontinent of faeces [...] Enodia is the name.[5]

And so on for three more types of seizure and three more gods. But this criticism also suggests that purifiers were offering a kind of empirically based diagnostic model in which the 'god' primarily served to classify the disease; treatment was predicated on this classification.

Moreover, as another of the author's arguments reveals, purifiers were advising clients to avoid or eradicate pollution by making dietary and lifestyle choices, such as avoiding goat products or wearing black.[6] Physicians who explained disease in naturalistic terms made similar recommendations, especially about diet. Thus the explanation of the mechanism (diet-caused bodily phlegm that produces a mechanical blockage in normal functioning, versus a diet-produced pollution of the person that causes a divine force to intervene in normal functioning) may vary, but the notion that effects and causes can be classified, understood and predicted, does not.

The interchange in trade and ideas between the societies and cultures around the Mediterranean basin, from Italy in the west to Babylonia, Persia and Egypt in the east, produced a steady influx of new recipes, ingredients, rituals and practitioners into the Greek and Roman mainlands. This exoticising diversification of practice and ideas was already visible in the period of the Hippocratic writers. As the conquests of Alexander and subsequently the expansion of Roman power made the world a smaller place, this process accelerated. By the time the elder Pliny (AD 23–79) wrote his encyclopaedia of *Natural History,* the traditional Roman medicine of Cato and the naturalistic medicine of Rome's Greek physicians were competing not only with the religious forms of healing offered by purifiers or by 'incubation' at the temples of Asclepius, but also with medical advice developed from practices of the magi of Persia, 'Chaldean' astronomers of Babylonia, Greco-Babylonian astrology and various other syncretisations and appropriations.[7]

It was not only these fashionable exotica that challenged Pliny's view of what medicine should be. Naturalistic, Hippocratic-style medicine had become the dominant paradigm among the Greco-Roman socio-political elite. The household medicine of Cato's Rome had been marginalised and romanticised as part of the Roman reception of Greek culture. Physicians in late Republican and imperial Rome were almost invariably Greek-speaking, like their primarily aristocratic clients: the cultured education of the upper classes included familiarity with the works and ideas of past and contemporary Greek physicians as part of natural philosophy. The aristocratic philosopher Calvisius Taurus (AD *c*.105–70), touring rural Greece with some of his friends, called in a local doctor after one of them had an accident and was surprised to discover that the doctor – unlike Calvisius Taurus and his well-educated social circle – did not know that there was a difference between veins and arteries, discovered in mid-third-century BC Alexandria.[8] In the Greek-speaking world that had resulted from Alexander's conquests and the advent of the successor kingdoms, cities seeking to guarantee access to reputable medical care for their citizens through the institution of the 'civic physician' recruited individuals from centres and teachers of Hippocratic-style medicine.

The Greek word for physician is *iatros*. In the earliest extant Greek literature, the eighth-century BC compositions the *Odyssey* and the *Iliad*, a *iatros* was usually a kind of craftsman. (The aristocratic hero sons of the medical god Asclepius also cared for their fellow-warriors with topical drugs and medical incantations.) Later physicians were still craftsmen–practitioners of a learnt expertise, a *technê*, like ship-builders or musicians. But the leading members of this group of professionals formed part of the Greco-Roman cultural elite and their discipline formed part of philosophy's intellectual revolution. Their dominance was contested and complemented by alternative, non-naturalistic forms of medicine, but was overall part of an extremely successful cultural meme. As we saw above, the emphasis on diet and elaborate diagnostic classes of the purifiers are reminiscent of Hippocratic practice. In Asclepiean temple medicine, patients dreamt of the god performing surgery, applying topical drugs and – especially in later sources – giving advice on dietary and exercise regimen.[9]

When to Use an Amulet

Some modern authors go so far as to describe the naturalistic approach as 'rational' while the common ancient belief in god-sent dreams is 'irrational'. But what do we mean by this, and is being rational the same as being scientific?

Take, as an example, the approach of the physician Galen to a popular 'alternative' remedy of his own time, an amulet packed with plant material. An amulet was usually worn on the body, ensuring that its effect was continuous (an ancient version of a time-release drug), but similar artefacts could be placed in the home. Their power for protection or cure was often attributed to the inscriptions and images inscribed on them, but several, like the one Galen examined, also used material substances with religious-magical properties, such as certain plants.[10] Galen tested one such herbal amulet and concluded that it was empirically effective, but not for the magical-religious reasons given by those who used or prescribed it. Instead, its effectiveness must be due, he argued, to 'effluvia', that is to invisible emanations given off by the plant's own substance. Absorbed by the wearer, these affected the balance of the body's humoural qualities, causing improvement in the patient's health. In a similar test of green jasper mineral for the health of the stomach, Galen found that jasper was effective by itself, without the amuletic design that religious-magical practitioners customarily engraved upon it.[11]

In the modern world, Galen's effluvia explanation, while wrong, seems to make more sense. Galen's take seems scientifically plausible to us because it rejects the therapeutic properties of amulets, just as modern biomedicine has no space for religious symbols or magnetic 'relaxing' stones beyond the placebo effect. Similarly, in our eyes, *On the Sacred Disease*'s 'natural' explanation for epilepsy – too much phlegm in the brain – places the blame in the right section of the body, identifies it as a problem of the organism within its environment rather than the act of a god, and detaches it from questions of blame, sin and pollution. The theory itself may be completely wrong, but the associated complex of attitudes is familiar.

On the other hand, not only were both theories in fact mistaken but there was very little evidence for either. This comment applies to much of ancient science, especially ancient medicine and biology. Much of its 'rationality', as opposed to the supposed irrationality of those who believed in the healing powers of amulets or the reprisals of gods, is an effect of hindsight. Science has revealed that on average plants are more likely to have a pharmaceutical effect than religious or magical artefacts are (even though many people now, as in antiquity, believe in the curative power of a religious symbol as well as in the science-backed chemist's prescription). We can credit Galen and his fellow 'rationalists' with backing the right horse in the long term; and perhaps one could argue that there was more evidence, consistency and logic in naturalistic explanations as a group than in the often conflicting rationales of rival beliefs.

The other aspect of Galen's reasoning here that we might identify as rational or even scientific, is how he says he arrived at his conclusions. He argued for the combination of empirical evidence, such as observing that a particular amulet works or carrying out anatomical dissection,[12] and logically grounded reasoning including *a priori* concepts, in order to arrive at conclusions about the unseen workings of the body that are epistemologically reliable and could be converted into actual treatments. In these broad terms, this is how we define the modern scientific method. But the difference is in the detail.

The idea of testing a treatment, like an amulet, to see if it actually works and, if it does so, if that success is due to one element of the treatment in particular, is a scientific one so far as it goes. We have no idea how Galen actually went about trying out the efficacy of the amulets with and without inscriptions, but it is probable that only one or a few subjects were involved. It is also rather convenient that Galen discovered empirical support for the conclusions he found acceptable in advance: that plant and material substances are likely to have natural therapeutic properties that affect the body by a physical mechanism. Finding what you expect to find is an eternal hazard in the scientific process, and part of the reason for the use of double-blind medical trials. (Galen's result here is particularly good news for him, since it means he can allow or even prescribe treatments that were popular with patients, without having to surrender his scientific views on what should work and how it does so.)

The most complex and careful 'experiment' in medicine known from antiquity was carried out by the physician Erasistratus of Ceos (*c.*304– 260 BC), probably working in Alexandria or another Greek-run Near

Eastern capital, perhaps Antioch. Erasistratus came from the Aristotelian tradition of empirical investigation, then at its zenith in medicine. His fame results largely from his application of dissection, which had brought Aristotle significant rewards in the understanding of biology, to human anatomy, physiology and pathology.

The idea he was attempting to test was that humans and other animals lose fleshy material in the form of invisible airy emanations, what we would call gases, through pores in the skin. The existence of sweat indicated that such tiny ducts existed and the idea that breathing, both through respiration and exhalation, takes place partly through the skin, had been a medical theory since the fifth century BC.

According to an account by an anonymous medical history, an excerpt of which is preserved on a papyrus known as *Anonymus Londinensis*, Erasistratus tested this idea in the following way:

> If one were to take a creature, such as a bird ... and were to place it in a pot for some time without giving it any food, and then weigh it with the excrement that has visibly been passed, he will find that there has been a great loss of weight, plainly because, perceptible only to reason, a copious emanation has taken place.[13]

Erasistratus weighed a captive bird at the start of the experiment and then kept the bird imprisoned in a small cage without food or water for some days, measuring the weight of all excreta from the bird during that period. At the end of the experiment he weighed the bird again, which had naturally lost weight, added this figure to the total of the bird's excreta and compared the result to the bird's original weight. If the original figure were more than the bird's final weight plus the weight of what it had visibly lost over the period, then the discrepancy would indicate that some of its original mass had been lost in a non-visible form, presumably that of the hypothetical gaseous emanations. This is indeed what Erasistratus found. In the seventeenth century Santorio Sanctorius carried out similar experiments on himself, with the same result.

This is a procedure that meets many key criteria for a scientific experiment, especially if we take some elements unrecorded by our sources on trust, such as what instrument Erasistratus used for weighing and whether it was in fact always the same instrument, calibrated in the same way. The procedure is repeatable, albeit unrepeated (so far as we know). It was intended and designed to prove or disprove a single hypothesis and to this end controlled and excluded as many variables as possible. It was quantifiable: weight was one of the few metrics available to antiquity.[14] To the extent that the theory predicted an experimental result, it is arguably even an instance of a crude form of Popperian falsifiability, since finding no unaccounted for weight discrepancy would have counted against the skin-exhalation idea.

The phenomenon itself, as observed by Erasistratus and Sanctorius, is real. Body weight fluctuates constantly over quite short periods, such as a day or night, as a consequence of metabolism. In respiration, cell mitochondria oxidise glucose or other sugars. Byproducts of this process are carbon dioxide and water, which are then exhaled from the lungs through the mouth – the latter in the form of water vapour. The amount of carbon dioxide is in weight terms small and can be matched or surpassed by the weight of oxygen, but the water vapour totals are a more significant amount. A human can lose 150–500 grams this way during a 24-hour period. Energy is also lost from the body as heat and as the energetic costs needed for the chemical reactions.

As it stands, then, the experiment was sufficient to demonstrate that something needed explaining, but could not sufficiently narrow down the options for that explanation. Erasistratus had no means for measuring the products and amounts of what the bird was exhaling. More importantly, there was, at this time, insufficient understanding of the chemistry of respiration and digestion to produce realistic hypotheses as to what the cause of the invisible weight loss might be. The skin-breathing theory of respiration was the only one available to account for the demonstrable loss in weight, but taking the latter as sufficient confirmation of the former fails to realise that other and more complex physiological mechanisms might produce the same result.

QED?

Another such classic of experimentation is Galen's demonstration of the role of the kidneys in the excretion of urine in an effort to disprove a rival theory, put forward originally by the physician Asclepiades of Bithynia (*c*.124/9–90 BC), that urine came from the blood directly into the bladder:

> Now the method of demonstration is as follows. One has to divide the peritoneum in front of the ureters, then secure these with ligatures, and next, having bandaged up the animal, let him go (for he will not continue to urinate). After this one loosens the external bandages and shows the bladder empty and the ureters quite full and distended – in fact almost on the point of rupturing; on removing the ligature from them, one then plainly sees the bladder becoming filled with urine. When this has been made quite clear, then, before the animal urinates, one has to tie a ligature round his penis and then to squeeze the bladder all over; still nothing goes back through the ureters to the kidneys. Here, then, it becomes obvious that not only in a dead animal, but in one which is still living, the ureters are prevented from receiving back the urine from the bladder. These observations having been made, one now loosens the ligature from the animal's penis and allows him to urinate, then again ligatures one of the ureters and leaves the other to discharge into the bladder. Allowing, then, some time to elapse, one now demonstrates that the ureter which was ligatured is obviously full and distended on the side next to the kidneys, while the other one – that from which the ligature had been taken – is itself flaccid, but has filled the bladder with urine. Then, again, one must divide the full ureter, and demonstrate how the urine spurts out of it, like blood in the operation of venesection; and after this one cuts through the other also, and both being thus divided, one bandages up the animal externally. Then when enough time seems to have elapsed, one takes off the bandages; the bladder will now be found empty, and the

whole region between the intestines and the peritoneum full of urine, as if the animal were suffering from dropsy. Now, if anyone will but test this for himself on an animal, I think he will strongly condemn the rashness of Asclepiades, and if he also learns the reason why nothing regurgitates from the bladder into the ureters, I think he will be persuaded by this also of the forethought and art shown by Nature in relation to animals.[15]

This careful set of procedures methodically piles one piece of evidence against Asclepiades on top of another, proving the role of the kidneys in urination by showing that they are full of urine, that the flow is one-way, that if allowed to function normally a ureter discharges into the bladder, that the pressure in a full ureter is high, and that disabling both ureters means that the bladder does not work and a pathological state is induced.

The success of this proof-by-experiment is partly because Galen is answering a simple question about which anatomical structures are involved in a physiological process. Experimental evidence, even when it relies upon apparently verifiable and empirical anatomical dissection, is more vulnerable when it tries to answer more complex questions about the mechanism by which such processes happen. Galen performed a set of incisions in the oesophagus, showing that none of these prevented swallowing. This showed, he said, that 'this cannot possibly take place in any other way than by the stomach drawing the food to itself by means of the gullet, as though by a hand'.[16]

What he means by this curious remark is that the stomach attracts food to it, in just as definite a way as it would if it literally stretched up and dragged the food down the throat. Galen thinks that his experiments prove that the swallowing motion of the throat is not transferred down the throat from the larynx or head, but is subsidiary to the stomach's attractive power. This theory is just one example of a larger overarching interpretation in ancient medicine about how anatomy and physiology intersect, that is through what Galen called 'natural faculties' or 'natural powers'. Each organ naturally has powers of attraction, assimilation, expulsion and growth; although what might be attracted or assimilated varies suitably from organ to organ. The idea did not originate with Galen and many other influential physicians and philosophers subscribed to some version of it.

It may be easier to understand by thinking of the attractive property of magnetic stone, of which the Greeks and Romans were well aware. The notion of the body's having 'natural faculties' is essentially that their shape, material and placement creates an organ that, as more than the sum of its parts, exerts a precise and particular effect on their environment, just as the magnet has the power to attract iron. In the case of the stomach, for example, it drags food down to it (in combination with the nature of the oesophagus). Similar reasoning explained the heart's power of beating, the bladder's ability to attract urine and the pulse felt in an arterial wall.

This is not a mechanical system. It is not simply because the stomach is lower than the mouth that fluid falls into it. Pressure does not force air or blood or urine through vessels. Such theories existed and influential versions of them had been articulated by Erasistratus and Asclepiades. Galen was vigorously opposed to both. The theory of natural faculties is about the power of nature as expressed in its details and is closely related to the teleological functionalism of Aristotle. It is as if the organs of the body were a jigsaw puzzle designed by nature to fit together so that each segment achieves its purpose as part of the overall picture of what it is to be a functioning human being (or dog, or plant, with different parts as necessary). If you replaced the stomach with a plastic replica or a second heart, it would not work: but what a proper stomach does naturally is be filled with food and what the oesophagus does naturally is swallow, just as a certain kind of flower naturally attracts a bee or an object naturally falls downwards, grasped by the ground. On this view, no further explanation is required: the question of why the bladder attracts urine is answered by the fact that it does – as experiment demonstrates – and so urine-attracting bladders are a jigsaw piece in how the universe works. The investigator unsatisfied by this answer, and just-so theories in general, has to enquire further.

Galen and Erasistratus clashed on precisely this issue, over the question of what causes pulsation in the arteries. By the time that Galen was learning medicine Erasistratus had been dead for over 400 years. However contemporary physicians, the Erasistrateans, still defended or modified his ideas and practiced on their theoretical basis. In attacking Erasistratean theory Galen was arguing against both Erasistratus' own works, as his treatises were at this point still in existence, and his modern defenders – Galen's rivals.

Erasistratus may have been the first person to distinguish between arteries and veins on an anatomical basis, although his slightly earlier contemporary Herophilus, who measured the arterial coat as being 12 times as thick as that of the vein, might have an even stronger claim. In any case, both did work in anatomy that transformed medical knowledge of the structures of the heart and venous system. Herophilus explained the pulse – found only in the arteries – as a natural and co-dependent capacity of the arteries and heart, as did Galen. Erasistratus, however, had offered a radically new idea. He argued that the heart functions as a bellows, forcing blood and air around the body through differentials in pressure. He correctly predicted that vessels too small to be seen connect the arterial and venous systems. He also believed that arteries under normal circumstances contain only air. In Erasistratean physiology, therefore, air (*pneuma*) is mechanically expelled from the left ventricle of the heart as it contracts, just as in a bellows or a pump. As the air is forced into the aorta the artery is forced to dilate: Erasistratus may have been aware that arterial walls are far more elastic than veins. The arterial wall rebounds, contracting due to this elasticity and pushing the wavefront of the air further down the artery, and the process continues.

There is one obvious problem here, from our point of view: arteries do not contain air (and neither does the heart).[17] Almost all Greek and Roman physicians and philosophers, including Erasistratus, thought that air in living beings, with its capacity to move faster and more easily than any liquid, was the carrier of perception and thought – the material that filled what we would call the nervous system. It was as necessary throughout the body as blood.

And although we no longer think of nervous messages as carried on an internal wind in this way, it is true that oxygenated blood, circulating throughout the body, is essentially the stuff of life. Erasistratus is not wrong to think air is vital and must be everywhere in the body: his mistake, as it turned out, was to think that it worked separately from blood. The very fact that the body contains two different kinds of vessels, running through its length and breadth, may have seemed to him an obvious sign of this division – different delivery systems (arteries and veins) for different substances (breath and blood) with different functions (communication and nourishment).

For our purpose the point is that Erasistratus viewed arterial pulsation as the result of fluid dynamics. Galen, on the other hand, thought it the result of a natural capacity of the heart working in concert with the natural powers of the arterial wall. The brain's ability to transmit motion along nerves to muscles, which Galen as well as Erasistratus had demonstrated in vivisection, was an analogical ability.

Erasistratus was said to have demonstrated mechanical pulsation in the arteries by means of animal vivisection. Galen, carrying out the same experiment, says that the result actually disproves a mechanical pulse, and will 'end the argument'. He describes his procedure in some detail:

> It is this. If you will expose any one of the large and obvious arteries, freeing it first from the skin and then from the matter that lies over it and around it, so as to be able to put a ligature round it; then open it along its length and insert a hollow reed or small bronze tube into the artery through the opening, so as to mend the wound with it and prevent hemorrhage; then so long as you study it in this condition, you will see the whole artery pulsating. But when you put a ligature round and press the coat of the artery against the reed, you will no longer see the part beyond the ligature pulsating,

although the passage of the blood and the pneuma through the hollow of the reed proceeds as before to the end of the artery. If the arteries got their pulsation in this way, then even now pulsation would be continuing in the parts beyond the ligature near their ends. Since this did not happen, it is clear that the faculty that causes the movement of the arteries is transmitted to them from the heart through their coats.[18]

William Harvey tried the same procedure in the seventeenth century, but found it almost impossible to carry out because of all the blood. His replacement tube also seems to have fitted loosely. In spite of these difficulties, he found that the pulse continued below the ligatures and incision – as you would expect if the flow was indeed a matter of fluid dynamics through any tubular structure, but not if 'the power is transmitted along the walls' from the heart.

So the same experiment had a different result depending on the theoretical beliefs of the person proposing it: each individual found what they were expecting to find. It is not entirely clear how Galen failed to observe pulsation below the ligatures. Twentieth-century scientists trying to work out how he could have seen what he says he saw, using his own lengthy and precise description of experimental procedures as quoted above, could not get the experiment to not work (as it were). They suggest that Galen's substitute reed may have been crushed without his noticing, blocking the blood flow; and that he never in fact tried the procedure with the more resilient brass tube that his text also suggests.[19]

The case of the artery that didn't pulse shows several things clearly. Firstly, that it is not always enough that an experiment is repeatable in principle: it also needs to be repeated in practice and by different people if its conclusions are to prove robust. Secondly, that in antiquity such tests tended to act as additional and semi-rhetorical support for theoretical positions established on far more general grounds. They were not conclusive or widely used as tests. The evidence rather suggests that Galen's habit of demonstrating theory through public performances of vivisection was unique to him and his opponents attacked the results without repeating the methods.

Thirdly, it demonstrates a problem for the historian in evaluating whether some of the 'experiments' described in ancient texts actually took place or, if they were carried out, whether the results were adjusted in line with the author's expectations. Such appeals to empirical evidence are invariably introduced as a proof of whatever position the writer is advocating and rarely use the first person, instead describing the experiment or observation as an exercise for the reader and stating what the reader would find if he did so. Whilst some such procedures certainly were performed, particularly as a means of discovery in areas like anatomical investigation or mechanical invention, in many cases the author is merely expressing confidence in his prediction.[20] It is also frequently unclear whether the author of a particular textual passage, outlining such a proof-by-observation or proof-by-test, is describing their own work or a claim made by someone else as to what they have seen (or thought they would see).

In this particular instance, there is a chance that Galen did not perform this procedure or had problems carrying it out amidst the profusion of blood (as Harvey complained) and the difficulty of fitting the relatively fragile reed within the artery. He may have given up early or decided that any pulse below the artery was a problem with practice, not his theory.

It is also notable that even accurate and revelatory empirical investigation had only a limited impact on theoretical science and philosophy. When Herophilus and Erasistratus showed conclusively that thought and feeling were located in the head rather than the heart (since cutting the nerves that led to the brain abolished perceptual and motor function), the leading philosophers of the Stoic school at the time explicitly rejected it. Their own natural philosophy had already integrated a different anatomical model into their theories of thought, epistemology and language. Moreover, this anatomical model, in which we think and feel in the heart, was supported by poetic language, a traditional form of divinely inspired authority which the Stoics viewed as at least as reliable as the views of physicians. The latter were dismissed as unable to agree and inconsistent.

Thus the new anatomy did not cause a paradigm shift in ancient thought in general, although the location of thought in the head did slowly become established within ancient medicine. It was, crucially, accepted and confirmed by Galen, whose later authority guaranteed its orthodoxy from late antiquity onwards.

A more important issue is that improvements in anatomical understanding were not productive of better medical science. The work of Galen, Herophilus and Erasistratus demonstrated that empirical investigation, even when carried out to a very high standard and with reliable results, does not necessarily produce correct theory, let alone useful treatment. The anatomical investigators of antiquity were able to describe the heart and vascular system, digestive and urinary organs, reproductive structures, nervous connections to the brain and anatomy of the eye in, mostly, unprecedented detail and accuracy. (Herophilus and Erasistratus were aided in this by being able to work on humans both living and dead: see Chapter VI.)

As a result, they achieved an unusual level of consensus about the basic map of the human and mammalian body. This did not translate, however, into consensus about how that body worked, what happened to it in cases of illness, in what disease consisted or about how best to help. There was disagreement even between anatomical experts about what flowed through the arteries, what caused the pulse, whether menstruation was good or bad for female health, how digestion worked, if blood-letting was a valid treatment. Differences were not only over specifics: the 'natural faculties' preferred by Herophilus and Galen had nothing in common with the mechanical conception worked out by Erasistratus.

Scholars have suggested that some contemporaries also perceived anatomical investigation as a failed project and consequently turned away from such methods.[21] The 'Empiricists', a group of doctors with a common methodological approach, abandoned all theories into the nature and causes of illness, health and human bodies as unknowable. Their founding member was a former pupil of Herophilus. A couple of centuries later, the 'Methodical' school of medical thought adapted the speculations of a popular physician, Asclepiades of Bithynia, to produce a simple rubric for diagnosis and treatment, devaluing anatomy and physiological theory as not unknowable but irrelevant. Physicians from both these groups became highly successful. Much of what Galen writes is in reaction to the 'Empiricists' and 'Methodists', making an assertion of the case for medical knowledge like anatomy against this common view about its irrelevance to the treatment of the sick. Good anatomists were better surgeons, as Galen points out, but it is unclear if patients were listening to him. In other ways the disputes between physicians must have seemed arcane, particularly as the treatments given to patients varied less than the explanations as to why they worked. The example of the amulet, which we began this section with, shows this clearly.

The Contagion Superstition

In 430 BC the city of Athens and its nearby port, the Piraeus harbour, fell victim to epidemic disease while already under siege from a Spartan invasion. The episode is described by the historian Thucydides in his unfinished account of the Peloponnesian war between Athens and Sparta.[22] In fact, he is our only contemporary source for the event – no other surviving text from the period mentions the epidemic of Athens.

Thucydides was an eyewitness and suffered himself from the disease: unusually, he recovered. His account is a classic of description, conveying the horror of the outbreak and its catastrophic effects upon the morale of the inhabitants as a microcosm of the war. It shows signs of Thucydides having read contemporary medical texts. He is familiar with the terminology and his tone is clinical, detailed and apparently objective, sounding like that of the case histories in the medical treatise *Epidemics*.

He reports both that some people thought the epidemic began with the Spartans poisoning the Piraeus' water supply and the alternative view, that the plague began in Ethiopia and then came north; but is himself non-committal. He notes that dogs and birds died after eating the corpses of those who died from the illness, and that those individuals who recovered from the disease, like himself, did not become ill again. People, he says, began to abandon those who were ill because contact with them was so hazardous. Physicians died in the greatest numbers because they visited the sick most often (2.47.4) and people were terrified at the sight of people dying like sheep after they had caught the disease from others (2.51.4).

This is an interesting remark, particularly as Thucydides' phrasing suggests he thought the fear a realistic one. The idea that illness could be passed on from one person to another through proximity or contact would seem to be a matter of simple observation. Greco-Roman medical texts, however, allow for person-to-person transmission only as a subset of an environmental cause of disease, and for a relatively small number of illnesses.[23] Historians have consequently supposed that most physicians, from at least the fifth century BC on, equated contagion with the religion-associated idea of pollution.

Illness marks a person out in a negative way. It is a mark of disfavour from the gods or, at best, a demonstration of ill luck. In Greek myth, tragedy and Herodotus' *History*, retribution for impiety or the breaking of social taboos often takes the form of physical or mental illness, particularly blindness or madness. Seizures were interpreted as acts of the gods and described as the 'sacred disease'.

Not all such cases were due to the dramatic mayhem found in myths. Menstruation, childbirth, sexual intercourse and death separated the pure from the polluted on a temporary basis. Illness also expressed more everyday forms of pollution: some of them merely the inevitable consequences of living in an impure world. The cause of the pollution might not be obvious, but the experience of illness suggested that something about oneself was out of joint with the world. It had to be cured by appeal to the gods, through the knowledge of purifiers or by the avoidance of polluting things (such as goat-skins, according to the purifiers) and the use of appropriate symbols, plants, animal products and words. The recovery from illness, physical or mental, marked the person's reintegration into the world, the favour of gods, civic society and their daily life of family and work. It normalised them.

Associating with the ill was dangerous: pollution was transferable through contact, proximity, acceptance. One effect of this belief was that persons and families suffering from illness were socially excluded on at least a temporary basis. This would have had some effect at limiting the spread of infection. The concept of pollution was also flexible enough to provide a means of understanding inherited disease within families; not only in ancient Greece but across all kinds of societies and surviving in English phrases like 'bad blood' or 'tainted bloodline', with their pejorative connotations.

The naturalistic worldview, however, seems to have excluded pollution. There is no particular text which focuses on arguments against it, but the author of *On the Sacred Disease* indirectly criticises this kind of metaphysical causation in his criticisms of the purifiers and others who argued that the sacred disease – epilepsy – was the result of contact with polluting substances.

The invisible mechanism of contagion itself seems likely to have been a problem for naturalistic thinkers. It suggests that illness transfers from one individual to another by magic, without any visible exchange of material. They needed another explanatory framework for disease, and it needed to be one which could account for people getting sick after close contact with those already ill. Their solution was the notion of environmental cause.

The leading theories of physiology and disease in antiquity were the humoural ones, according to which the human constitution consists of a small number of fluid substances (humours). Although, as we have seen, this was not the only theory of disease circulating in antiquity it is true to say that humoural theories dominated, particularly in the four-humour version (blood, yellow bile, black bile and phlegm) articulated by *On the Nature of Man* and later taken up by Galen.

In a healthy person, humours and their particular qualities of hotness, dryness, wetness, and coldness – phlegm, for example, is wet and cold – are in balance. Illness is the result of imbalance. Medicine restores humoural ratios to a healthy or normative state or prevents an imbalance from occurring. Inherited or environmentally caused variations in which a person has more of one humour than another explain differences in body type, inherited conditions and predispositions to particular illnesses. The linkage of humours with temperament survives in modern English: people are phlegmatic, sanguine, melancholic, or choleric, from the Greek or Latin words for phlegm, blood, black bile and bile (*cholê*).

The body's humours and hence the body's dryness or wetness or coldness or hotness were affected by all kinds of interaction with the environment, from what people ate and drank to the air they breathed in and the prevailing temperature of the winds and climate. The fourth-century BC medical treatise *Airs, Waters, Places* describes, with considerable attention to detail, how local environment produced different bodily and mental types:

> Let us now take the case of a district [...] sheltered from the south but with cold prevailing winds from the quarter between the north-west and north-east. The water supply is hard and cold and usually brackish. The inhabitants will therefore be sturdy and lean, tend to constipation [...] more troubled with bile than with phlegm [...] the special diseases of the locality will be pleurisy and acute diseases [fevers] [...] Those under thirty suffer nosebleeds which are serious in summer [...] Characters are fierce rather than tame.[24]

Clusters of a particular kind of illness could be explained through such localised environmental variation. Winds bringing air from hot and dry climes affected neighbourhoods or whole villages and cities, explaining epidemics. Conversely, even over a relatively small area there can be variation in exposure to wind, rainfall, soil type and water drainage; families are likely to eat the same foods (grown in the same soil and water). Congenital, inherited characteristics of bodily constitution, meanwhile, mean that not everyone exposed to the same environmental factors reacts in quite the same way. That explains how one person might show signs of illness before their wife or neighbour, a pattern that could otherwise be mistaken for contagion.

This is a wonderfully flexible theory. Thucydides' remarks might suggest that it was not entirely successful in reinterpreting the evidence for contagion in the teeth of common experience, but no other idea gained much traction over the rest of antiquity. The nearest equivalent to a theory of infection was a variation on the environmental theme: *miasma*, or bad air, which developed later into an idea of small airborne 'seeds' of disease.

Miasma is the Greek word for what I have translated as 'pollution': a kind of atmosphere of wrongness which can also be read as the literal atmosphere. Air is a good candidate for a vehicle of disease, as it gets everywhere, is unavoidable, and is both rapidly mobile over long distances and a feature of any local climate.

Other kinds of bad air might be very local. Sufferers from certain diseases, say later medical authors, breathe out or give off a kind of poisonous air. Their skin, too, might produce toxic excretions that either poisoned the surrounding air or could be absorbed by touch. These excretions, if breathed in or otherwise absorbed by someone with a humoural predisposition to that illness or an unhealthy dietary regimen, were often enough to tip that person's balance of qualities over into the same kind of physiological abnormality. It made taking care of people with such disease a risky proposition – just as with the Athenian epidemic.

Such toxic excretions were the kind of noxious air produced by rotting corpses – often found in association with a rapid spread of disease. Other sources of *miasma* were marshes and stagnant water. Gases can be seen rising from warm swamps, accompanied by swarms of biting midges and malarial mosquitoes. Malaria is from the Latin for 'bad air'. The Roman era architect and town planner Marcus Vitruvius Pollo (active *c*.40–15 BC) described the risks of such sites in his treatise *On Architecture*:

> For when the morning breezes blow toward the town at sunrise, if they bring with them mist from marshes and, mingled with the mist,

the poisonous breath of creatures of the marshes to be wafted into the bodies of the inhabitants, they will make the site unhealthy.[25]

The toxicity of an ingested substance is an obvious way to explain disease, being familiar from food poisoning (and other kinds of poisoning). In the second century AD Galen remarked that a drop of saliva from a rabid dog was enough to cause the disease in humans.[26] In the passage from Vitruvius, the concept of toxicity is stretched, through a set of associations between unhealthy cities and poisonous animals, into a direct causal link of material substances. Vitruvius and his sources could have been thinking of small biting insects – mosquitoes, the actual vector of malaria – among these toxic marsh residents. Alternatively, he might have had animals like water snakes in mind.

A Roman author on agriculture, Marcus Terentius Varro (116–27 BC) had already suggested that tiny animals, invisible to the eye, were breathed in and caused disease.[27] That kind of notion would only have been reinforced by the folk and scientific belief that insects germinated spontaneously from hot damp areas like marshes or decomposing animals (Chapter II).[28] Other texts explain the bad air of marshes as due to gaseous effluvia from such decomposition. Again, an association of ideas, this time between putrefaction and the spread of disease, has been actualised into a theory about the transfer of disease-causing matter. Even Galen made use of the notion of 'seeds of disease' in several works, as a way of explaining individual variation in response to common environmental factors.[29]

In antiquity, then, the idea of contagion was valid in naturalistic terms when conceived of as a substantive transfer of something already toxic, including insects born from decomposition or bodily fluids from sick animals or humans. The toxic substance was ingested or breathed in. But it remained only a contested hypothesis for a subset of illnesses, such as an oozing skin disease; or an extra hazard of a specific environment, such as marshy places. The dominant paradigm among physicians and intellectuals remained the idea of keeping the body in balance by adjusting to food and water intake and environmental changes, a goal mainly achieved through dietary and exercise regimens, supplemented in cases of illness by purgative or drying drugs.

The concept of seeds of disease was revived and elaborated in the sixteenth century by Girolamo Fracastoro, but they remained an entirely theoretical entity until the improvements to microscopes made over 100 years later by the Dutchman Anton van Leeuwenhoek, the first person to see actual micro-organisms. This immediately made 'seed' or 'germ' theory a much more likely possibility, given that its central hypothesis had now been proved to exist. The experiments of Louis Pasteur and subsequently Robert Koch, leading figures in the dawn of microbiology, confirmed that it was such micro-organisms that grew in food and fermentations and could be readily transferred by contact, and disproved the idea of spontaneous generation.

It is not surprising that the idea of contagion through very small organisms, while essentially correct, could not command general support until technological advances proved its central supposition. And without that, there seemed to be no mechanism by which disease could be transferred from place to place or person to person. Instead, contagion was assimilated to its traditional explanation, pollution, and explained away.

Humoural or other 'in balance' theories of illness, in their flexibility, were too successful as an explanation. They provided an expansive role for the physician, in both prevention and treatment, as the expert who could manipulate the sufferer into equilibrium with their environment: the physical version of the purifying restoration favoured in more traditional explanations. They took away some of the questions as to how illness actually behaved.

Thucydides tells us that it seemed to some of the people in Athens in 430 BC that the plague could be caught by contact with those already ill. He mentions a rumor that the illness originated in poisoned wells. It was technically feasible for either or both of these commonplace beliefs to be tested, as John Snow (with assistance) did in the case of the cholera epidemic in London in 1854, by showing that it originated in the contaminated Broad Street water pump and not in miasmatic air. A few years earlier, in 1847, Ignaz Sammelweis had realised that pregnant women whose childbirths were attended by doctors in his Viennese hospital were much more likely to develop puerperal fever than those whose deliveries were carried out by midwives. Investigating further, he found that the physicians often came to the childbirth subsequently to performing an autopsy on the hospital's dead, whereas the midwives had not. He instituted a system in which the physicians washed their hands in chlorinated water before examining pregnant women and cases of puerperal fever dropped dramatically.

As a leading exponent of naturalistic medicine, and its historically most influential advocate, Galen argued for the need to combine reasoned theory with empiricism. In the long run, we can agree: this is how we do biomedicine now. But the importance given to a persuasive theory, to the perception that we already understand what is happening, can also blind us to other interpretations. The work done by Snow and Sammelweis did not mark a shift in paradigm to a theory of micro-organisms. Instead, it was largely rejected and even reviled by other physicians.

This is not merely the unfortunate result of an underdeveloped nineteenth-century scientific culture. Take, for example, the peptic ulcers and gastritis which were believed throughout most of the twentieth century to be the result of stress, bacteria being thought unable to survive in the acidic confines of the stomach. They were treated accordingly, without much success. It was not until the 1980s that Barry Marshall and Robin Warren were able to demonstrate the causal role of the bacterium *H. pylori* – and Marshall had to resort to deliberately infecting himself to get their ideas and evidence taken seriously. This is a rather nice analogy to our overarching theme, since essentially this replaced a lifestyle and inheritance theory of disease with one of micro-organism infection, curable with a course of antibiotics.

In this case Marshall did in the end have the methodological resources and experimental data to prove his argument. Similarly microscopic evidence, Pasteur's experiments and finally Joseph Lister's work in the uses of hygiene against contagion eventually vindicated the conclusions of Snow and Sammelweis, among others. Antiquity, in contrast, lacked the technical resources, the developing experimental culture and the rapidly evolving sister field of chemistry that have been enjoyed by medicine over the last few centuries. The Greco-Roman version of the contagion hypothesis put more emphasis on transmission through the air rather than transfer by person-to-person contact, and this might have made it less likely that physicians would think of washing off potential contaminants.

Theories about Women

Another notable feature of ancient medical culture was its concern with the socially significant arena of female health and reproduction.

'The female is a deformed male.' Aristotle is analyzing reproduction and inheritance in his biological treatise *The Generation of Animals*. He is starting from the view that every animal, including man, is designed by nature and fitted perfectly to its purpose. But for most animal species each one comes in two versions: female and male, which have significant anatomical, physiological and – the Greeks thought – behavioural differences between them. If one version is perfect for what that species is and does, the other must be an inferior version.

For Aristotle, the male is the well-designed version. The female of the species exists for a different reason: a necessary but much more limited one than that of the male. She ensures that the species is able to reproduce and therefore continue, because she is designed by nature to carry and supply nourishment to the young of her species. The compromises that this entails in the original design means that a female goat or a female human cannot be as good a goat or as good a human as the male. The female, effectively, is a deformed male: deformed for the specific purpose of being able to bear children.

In Greek biology and medicine, mental differences – thinking, emotions, behaviour – were determined by physical differences. Differences in temperament, including excitability or a tendency to sadness, were caused by differences in the balance between the humours. Children had different capabilities and emotional characteristics to adults. Anatomical and physical differences between the sexes, therefore, were naturally thought to produce differences in male and female cognition and emotion. This was a case of science accepting an almost universal set of cultural beliefs about men and women. Physicians and other intellectuals supplied a theoretical grounding for these beliefs by explaining them in terms of animal and human anatomy and physiology. In medicine, ideas as to what constituted female health were defined largely in terms of a woman's ability to fulfil her socially validated roles, since cultural preferences were usually taken to represent a natural inevitability.

Aristotle listed the differences he thought he had observed in male and female behaviour:

> All females are less spirited than the males, except the bear and the leopard: in these the female is held to be braver. But in the other kinds the females are softer, more vicious, less straightforward, more impetuous, more attentive to the feeding of the young, while the males on the contrary are more spirited, wilder, more straightforward, less cunning. [...] even among cephalopods, when the cuttlefish has been struck by the trident the male comes to the female's help, whereas the female runs away when the male has been struck.[30]

For Aristotle, the human design produces rationality, our distinguishing characteristic as a species and our greatest and most necessary purpose. Since females are compromised human beings, so is their reason. As a result, in his ethical and political works, men have authority over women who, like children and slaves, are not really capable of managing themselves. Not only are they less good at reasoning but they have less self-control, making them moral weaklings.

Even in antiquity, Aristotle had one of the more extreme conceptions of female capacities and roles. (Nor did everyone agree with him about the biological inferiority of non-Greeks as a justification for slavery.) A different emphasis can be found in a treatise, the *Oeconomicus*, by a near contemporary of Aristotle, the politician and professional soldier Xenophon (431–355 BC). The *Oeconomicus* is about educating the ideal wife from a young age. Xenophon's paragon of female virtue is capable of excellent organisation in household management and is in general better than men at indoor activities. She has 'more affection for newborns'; she is more fearful because fear is a good quality in someone who has to manage the household stores. (Presumably it encourages forethought.) A man is more equipped to deal with military campaigns and physical danger, being more courageous and with greater physical endurance.[31]

Xenophon is, for a Greek or Roman author, unusually generous about women's capabilities. Men, he says, are no better in regard to memory or dedication or even in self-control. He represents anxiety – a negative for a man – as a positive for a woman. One of his characters observes that the wife being described has a 'masculine intelligence'. But if Xenophon emphasises complementarity rather than inferiority, he is clear that men and women have different capabilities. A woman's role is in household management (the literal meaning of *Oeconomicus*, 'economist') and caring for others, including children and slaves. Men are suited to farming, sailing, fighting: physical activities in the open air.

This set of differences correlated with physical ones. Women had less strength; they also had a different type of flesh. The writers of the fourth- and fifth-century BC medical texts that refer to female physiology describe women as being soft and porous, like a sponge. This natural characteristic could be reduced or intensified by their lifestyle: those women who did a lot of physical work (and of necessity many rural or poor women did just that) could harden and dry out their flesh until it was almost manlike. In the same way men who behaved like women became flabbier and damper. Female and male were not entirely fixed traits: behaviour had to be matched to nature or that nature became distorted. The writers of the medical texts make it clear that such mannish women and effeminate men are not ideal.

Women were not only damper and spongier than men, but hotter. That at any rate was the view of the medical texts on gynaecology. Aristotle, on the other hand, regarded them as colder than the male. This had crucial effects on their physiology, especially in regard to reproduction. In both sexes, Aristotle argued, heat enabled men and women to convert food into blood, literally cooking what they ingested into a different and usable form of nutrition. It was only men, however, who were hot enough to cook that blood even further and produce semen.

Semen, according to Aristotle, was a special kind of material. Much more special than blood. Only semen was complex enough to carry the information of inheritance from father to child, imprinting his form on the base material provided by the mother (her menstrual blood). The greater heat of the male was a vital part of his biological superiority. In theory, if one could turn down this thermostat, the result would not be male but something more like an anatomically divergent female; or as Aristotle preferred to think of it, a female is an infertile (deformed) male.[32]

The rather obvious flaw in Aristotle's theory of conception is that it makes inheritance paternal only. It cannot account for children who resemble their mother or someone else in their mother's family. (Aristotle's explanation for this, involving 'movements' in the mother's blood that can cancel out the paternal semen, is not his finest hour.) A very different theory is argued by the author of the gynaecological treatise(s) *On the Seed* and *On the Nature of the Child*. He subscribed to pangenesis, an idea also articulated by the Pre-Socratic atomic theorist Democritus. In this version of pangenesis, seed comes from every part of both parents' bodies. Seed from bone is the seed of new bone, that from the heart is the seed of the child's heart, and so on. Depending on the precise mixture of the two kinds of semen that make it into the mother's womb, the child will show the characteristics of whichever parent the majority of a specific body part came from. If, for instance, it was mainly from the father that seed from the eyes came from, then the child will have the same eye colour and shape as his or her father.

As for the sex of the child itself, *On the Seed* has an interesting answer: 'In fact both partners alike contain both male and female sperm (the male being stronger than the female must of course originate from a stronger sperm).'³³

If both partners produce male, that is, 'stronger', seed, then the child is male. If both supply female, or 'weak', seed, then she is a girl. It is however possible for the male to produce weak seed and the female strong seed, or the male strong seed and the female weak seed. In either of those cases the seed with the most quantity wins, whether it is strong or weak seed and whichever parent it comes from. The author, it turns out, is thinking of a kind of cake mixture in which the smaller amount of seed is simply lost in the mix:

> It is just as though one were to mix together beeswax with suet, using a larger quantity of suet than of the beeswax, and melt them together over a fire [...] only after it solidifies can it be seen that the suet prevails quantitatively over the wax. And it is just the same with the male and female forms of sperm.

In *On the Seed*'s terminology, seed which produces males is 'strong'; girls come from 'weak' seed. The seed itself does not seem to be strong or weak vis-à-vis its alternative form, although the author does not explain what happens in a mix of exactly 50 per cent each. The terminology seems to be based on results – weak sperm causes girls, who are physically weaker than boys – but it also clearly reflects the gender hierarchy of Greco-Roman society. Strong is to be preferred to weak. Similarly loaded cultural assumptions are visible in the twenty-first century: the phrase 'don't be a girl' means 'don't behave in a (stereotypically) female way' with implications of weakness and fear. This is the case even when it is said by a woman or directed at one; it is not always said with irony. 'Just like a guy' is similarly a mild criticism of stereotypically male behaviour, implying recklessness, aggression, stoical silence and empathy failure.

In another medical text of roughly the same period (*On Regimen*), this mix of semen from the parents correlates precisely to gendered behaviour along a spectrum. If the semen from both parents is female, the children are 'very female and very well-behaved'. If the women's seed is female and the man's is male, and the female seed dominates, then the girls are bolder than the girls produced from doubly female seed, but still appropriately modest. If this is reversed, so that the seed from the father was female and from the mother, male; then the girls that result are bolder than the others 'and are called mannish'. Similarly, boys can be intelligent, strong men (male semen from both parents), less so but still brave (male seed from father, female from mother) – or are 'men-women' (female from father, male from mother). Summing up, the author says: 'the degree of manliness depends upon mixing, and upon nourishment, education, and lifestyle'.[34] Behaviour is here not solely a matter of inheritance, since education and habits of life can go some way towards altering original tendencies, but certainly physiology is fate. Part of what that biological fate determines was where one fell on a gendered spectrum of the behaviour constructed by society as appropriate to, and natural to, one's sex.

Among the other characteristics thought to vary between women and men was their desire for and pleasure in sexual intercourse itself. According to *The Seed*, 'the pleasure experienced by the woman during intercourse is considerably less than the man's, although it lasts longer'; it is continuous throughout intercourse and it ceases when the man ejaculates. Other medical authorities, however, argued that women have as much or more pleasure in sex as men. They were also commonly thought to desire sex more than men: hence their lack of sexual self-control, a characteristic they shared with effeminate men. The plot of Aristophanes' anti-war comedy *Lysistrata* centres on a sex strike by the women of Greece, but they almost give way to their own desires several times.[35]

This cultural assumption extended to the belief that pregnancy could only result if women enjoyed sex. The medical explanation of this seems to have been that physical stimulation, in other words arousal, was necessary for conception. The Roman era physician Soranus was the author of the *Gynaecology*, a kind of handbook for what midwives or patients' families needed to know about female health, pregnancy and childbirth. Soranus accepts as medical fact this idea that there is no conception without female pleasure, but adds, citing cases in which women had become pregnant from rape, that the woman could be excited without realising it (drawing an analogy with hunger).[36]

We can see from these examples how medicine took its cue from widely held beliefs about the nature of men, women and sex; framed its inquires in those terms, and produced confirmatory justifications. These were not just the opinions of men. Although we are entirely reliant on male written sources, these do on several occasions cite the opinions of women as sources of information about female reproduction. One such female opinion was that an 'experienced' woman could learn to identify the moment of conception at the moment it happened, when the mixed seed did not fall back out but stayed within the womb as it closed up, triggered by the moisture.[37]

Identifying the precise date of conception usually involves some degree of guesswork, since pregnancy is not immediately obvious. Girls in Greco-Roman antiquity tended to marry relatively young, about 12–16 and in some cases younger. Some of them were probably undernourished (another medical text remarks that girls need less food than boys do). They may not have experienced menstruation before becoming pregnant: indeed the absence of menstruation in girls of about this age was frequently interpreted as a risk factor for illness or even illness itself and was 'cured' by sexual intercourse (in marriage). This did not make the task of establishing the date of conception any easier. Counting back from the day of childbirth is an obvious technique, but this was itself necessarily dependent on notions of how long pregnancy took – and estimating that involved identifying the moment of conception. It was in fact widely believed that a normal length of pregnancy was either nine months or, just as probably, seven or even ten months. An eight-month pregnancy was believed to be usually fatal for the child.

The author of *On the Nature of the Child* describes his own observation of an aborted foetus: a (female) relative of his owned a female slave as a dancer and prostitute:

It was important that this girl should not become pregnant and thereby lose her value. Now this girl had heard the sort of thing women say to each other – that when a women is going to conceive, the seed remains inside her and does not fall out [...] One day she noticed that the seed had not come out again. When I heard it, I told her to jump up and down, touching her buttocks with her heels at each leap. After she had done this no more than seven times, there was a noise, the seed fell out on the ground, and the girl looked at it in great surprise. It looked like this: it was as though someone had removed the shell from a raw egg, so that the fluid inside showed through the inner membrane – a reasonably good description of its appearance. It was round, and red; and within the membrane could be seen thick white fibers, surrounded by a thick red serum; while on the outer surface of the membrane were clots of blood. In the middle of the membrane was a small projection: it looked to me like an umbilicus, and I considered that it was through this that the embryo first breathed in and out. From it, the membrane stretched all around the seed. Such then was the six-day embryo that I saw. (13)

The author's interest in this episode and the care of his detailed description shows clearly how little information was available and how much guesswork involved in imagining the processes of conception and pregnancy.[38] What he describes is a good deal more developed than a six-day embryo.

At the time this treatise was written very little was known about female internal anatomy. Over 100 years later, in third-century BC Alexandria, Herophilus observed the Fallopian tubes and ovaries. He interpreted them on the basis of external male anatomy, as the female version of the spermatic ducts and testicles. It was an andocentric reading, but it made Herophilus sceptical of the widespread medical idea that there were diseases peculiar to women; a position with which Soranus subsequently agreed. In this case anatomical inquiry played a role in problematising well-established ideas about female health.

In texts of the fifth and fourth centuries BC, like *On the Seed* or *On Diseases of Women*, the health of women is exclusively tied to their reproductive status. Female social roles revolved around sex and childbearing, which explained their biological differences from men. Not having sex and children, therefore, was an unnatural behaviour that risked being unhealthy: 'Another point about women: if they have intercourse with men their health is better than if they do not.' (*On The Seed* 4).

Young women – adolescents – tended to suffer from *hysteria*, feelings of suffocation, pain and despair that were said to sometimes end in suicide. Such episodes were experienced by girls around the age of getting married: the traditional response involved rites of Artemis, the goddess of virginity and childbirth. The pre-marital female was a liminal figure, about to undergo the formal and actual process of maturation into a sexually active woman (the Greek word for woman is also the word for wife, *gyne*, from which we get gynaecological) and then into motherhood. In ancient society, only the virginity of cultic priestesses, a status available to few, offered any other route to social validation. Marriage, sex and childbirth was however both important and risky: cultural and individual anxieties surrounding this stage constructed the figure of the hysterical adolescent, out of control of her body and mind. Some girls would have felt it as reality: a known and almost accepted way of expressing the situation and their reaction to it.

Medical opinion, faced with this social and experiential phenomenon, constructed an elaborate explanatory framework. The term *hysteria* comes from the Greek word for womb. The earlier medical writers, lacking anatomical knowledge, thought of it as a mobile organ, light and hollow, easily displaced from its position against the cervix. It caused the symptoms of hysteria – pain, suffocation, mental disturbance – when it lodged against the heart or stifled breathing and voice in the chest. But this was only a risk in a sexually inactive or virgin body. Sex moistened the womb with a liberal application of semen, and the humid weight of that semen held it in place. The mixed seed developed into a foetus, nourished on a weighty supply of menstrual blood, making pregnancy even more effective than sex alone at preventing hysteria. Fortunately – in medical eyes – sex and pregnancy often went together, demonstrating the natural, normative, healthy status of the link between the two and their consequences for female health.

Menstruation, to many medical writers, was another sign of health. Women needed extra blood (which in Greek theory was converted from food and into flesh) not for the extra hard muscle of the male but for the material and nourishment of a foetus and (according to Aristotle), subsequently for breast milk. In a fertile women who was not pregnant, this blood had to be excreted as the menses. If the blood was blocked from exiting, problems would result.

So not menstruating (in the non-pregnant) became a symptom of ill health. It indicated that something had gone wrong with the reproductive mechanisms of natural female biology and hence with the female body in itself. The assumption was that the body was producing blood but not managing to excrete it. This might be because the blood was being stored up in the jar-like womb, from which it was unable to escape because the womb was misaligned with the vagina. Alternatively the mouth of the womb or the passage of the vagina itself was too fleshy and tightly closed. The stored-up blood could spill out of the womb at the top, causing physiological problems; or become toxic and putrefied, like food gone bad. Sometimes it might be excreted as nosebleeds – another symptom of hysteria. The cure was sexual intercourse (marriage), which would open up the passages and allow the blood to exit. Pregnancy was even more useful in that it used up the blood.

Not reproducing was dangerous because it interfered with the normal functioning of the female body. For the female, sexual or reproductive problems were necessarily whole-body and mental problems. The timing of the first menses was expected within a small window of time, and if girls were not menstruating by this age doctors recommended marriage (sex) as a matter of urgency, to relieve the build-up of entirely theoretical menstrual blood.

It was not a universal view. Soranus argued that virgins were quite capable of being healthy: indeed that not having sexual intercourse and the associated risks of pregnancy was for the woman the healthiest option. Virgins, he said, become not ill, but more like men. Society, he added, unfortunately requires reproduction for the survival of humans.

Soranus was an author much more sympathetic to the demands and capabilities of being female than was Aristotle, or many other medical authors. But in a way he thought of female and male similarly to Aristotle. Aristotle also said that the survival of the species was the reason for nature having to design the female, in a necessary compromise of the male template. In Soranus too, female biology is reproductive biology: take away that, and what remains is the male.

In the twenty-first century, anatomy has moved on. So has gynaecology. Society is certainly different. But culturally constructed ideas of adolescence, sexuality and gender roles remain. We have problematised but not solved their relationship to biology; nor have we worked out what either biological or environmental influences on our preferences, values and concept of self should mean for how we manage our society or ourselves.

Medical science is not a neutral observer in these debates. The science of contemporary biology is far superior to that of antiquity, but the certainty with which the latter's physicians articulated social constructs as obvious biological facts; or the ways in which its women thought they understood their own bodies – the moment of conception, the suffocating despair of virginity – is rather alarming. It makes one wonder what we are missing ourselves.

CHAPTER V

CONTROLLING THE WORLD

Rhomboids Are Forever

Contrasting with the hit-or-miss (largely miss) conclusions of the life sciences in antiquity is the history of mathematics and, to a lesser degree, of physics, engineering and astronomy. Mathematics simplifies the intricacies of physical-chemical or organic objects by abstracting and universalising essential aspects of them, so that a field becomes a triangle and a triangle is simply a relationship of three lengths and the angles between those lengths. In so far as the behaviour of real-world objects such as stars or projectiles can be abstracted into mathematical relationships, their movements and other characteristics can be predicted or even manipulated.

Mathematical techniques and theorems worked out in ancient Greece still hold true in the twenty-first century. The trigonometry of the theorem somewhat inaccurately associated with the Pre-Socratic philosopher-mystic Pythagoras of Samos (c.570–490 BC) is still taught in schools; just as geometry uses the same problems and proofs as those of Euclid's guide to mathematics, written in c.300 BC. Schools taught mathematics with Euclid's text well into the twentieth century and this area of math is known as Euclidean. (There is no evidence that Euclid himself was responsible for any of the proofs in his *Elements*: the use and perhaps the originality of his work lay in its systematic assemblage and arrangement of past and contemporary geometry, so that the reader began with a statement of the simplest, most basic assumptions and proofs. These were in turn used to prove more complex theorems, until the reader reached the most difficult and advanced point in Greek mathematics so far at the end of the book.)

The distinction between ancient and modern mathematics is therefore not disagreement but extent: algebra, differential or non-Euclidean mathematics, probability theory and much else have revolutionised the *reach* of mathematics, but not invalidated its early geometrical and arithmetical forms.

First Find Your Mathematician

The notion of an axiomatic-deductive proof itself is an invention of Greek classical culture. Other and much earlier civilisations used mathematical techniques: for example, Egypt's Rhind papyrus, which contains a mathematical text, dates from the middle of the second millennium BC. Pythagoras' theorem was applied in practice, long before Pythagoras formalised its structure, to the measurement of land by surveyors from the offices of the king in Egypt, for the purposes of taxation and flood management. Mathematics is also necessary for the design and building of structures like pyramids, palaces and temples. In Babylonia, a specialist group of priests observed and recorded the movements of the ominous stars with the aim of being able to forecast their future patterns of movement and adapt the machinery of state according to whether the omens for that day were good or bad. Similarly, arithmetic is vital to all centralised societies, so that they can count and measure their resources and determine the tribute to be paid to them. In the culture of Crete that was marked by use of the proto-Greek Linear A script, and subsequently Linear B, one of the chief uses of this early Greek literacy is to record the numbers of things (for example, how many sheep).

All these and other societies used maths as a set of rules of thumb, from case to case. Calculating field area by trigonometry was done using the equivalent of Pythagoras' theorem, but the calculation was always particular to the case. That is, you took the relevant lengths of the field in question and you put them together in the same way you always did and that you had been taught. This method always worked. It was therefore true in what we would call an inductive sense. Because it had always been true before, it was assumed that it would always be true again.

The concept of abstracting a mathematical technique and demonstrating that it must necessarily hold for any and all particular cases developed in Greek culture, in about the sixth century BC. This was the same period in which many of the Pre-Socratic thinkers were critiquing and refining common ideas of truth and knowledge and, in their startling speculations on the nature of the cosmos and its gods and purposes, trying out methods of proof and persuasion (Chapter I). It is in this context that the axiomatic-deductive mathematical proof was invented: a means of deducing the truth of a statement from inarguable first principles such as the definition of a line.

Applied to statements about non-geometrical objects, this became the origin of deductive logic. Where the modern world has learnt to prefer the empirically checked and largely inductive processes of science, many of antiquity's elites put their trust in formal logic as the means by which truth could be sorted from falsity. The syllogistic forms worked out by Aristotle in the fourth century BC remained a staple of education and argument throughout late antiquity and the medieval period and, like mathematics, they are not invalid modes of reasoning today, although their limitations are better understood. The Stoic school of philosophy made major contributions to the field shortly afterwards with the invention of propositional logic.

But the gold standard for deductive proof remained mathematics. Its success was demonstrated by the level of consensus among mathematicians themselves, which far surpassed any that could be found among doctors, politicians or philosophers. This is how Euclid was able to collect and organise mathematics into a single set of knowledge, whereas for any other subject such a collection would have summarised disparate and discordant opinions (as a doxography, that is, a history of thought). Plato, writing in the earlier fourth century BC, and Galen, some 500 years later in the second century AD, both use mathematics as an ideal analogy for the certainty of knowledge that they want their own subjects, respectively philosophy and medicine, to achieve.

Yet few people became mathematicians. The historian of ancient mathematics Reviel Netz has catalogued the names of 144 individuals for whom we have some evidence, often slight, that they did mathematics.[1] (The criterion is someone who wrote down an original mathematical demonstration.) They range from the later fifth century BC to the sixth century AD. Most but not all were male. They came from all over the Greek-speaking world and a high proportion, at least, were from the wealthy strata of society.

Netz extrapolated from this number and arrived at a total of 1,000 or so mathematically active people over about 1,000 years. Some periods may have had more than others. As he points out, the vast majority of inhabitants of the Greco-Roman world were non-participants in what we think of as Greek and Roman culture: they were illiterate, excluded, sometimes enslaved, often female, poor in both materials and time. But even in the quite small world of those not only culturally visible but counted among its literate, generalist, educated elites, most producers of culture were writers of literature and most of the audience did not read mathematics. The discipline was isolated and so were its practitioners.

Mathematicians belonged not to institutions, as they by and large do nowadays, nor to the doctrinal groupings of antiquity's 'schools' in medicine and philosophy, but within a loose network of individuals. In a period in which we know of more mathematicians than we do for other eras, Archimedes of Syracuse in Sicily (c.287–12 BC) corresponded with the astronomer-mathematician Conon (c.280–20 BC) while Conon was working in Alexandria and, subsequently, with Conon's pupil Dositheus (active c.250–200 BC). Archimedes' work 'On the method of mechanical theorems' was addressed to yet another resident of Alexandria, the polymath Eratosthenes (276–194 BC).[2] Such written correspondence was an important means of communication for most mathematicians, since they were geographically widely distributed.

This was not simply a matter of the best mathematicians being relatively rare or of a scarcity of experts in a particular branch. Netz's article makes a persuasive case that original mathematical work about difficult problems was not a popular field of study among the cultured, even when it was admired for its successes. This is similar to its perception in modern culture as well: beyond the dutiful, enforced learning of multiplication tables and Pythagoras' theorem in school, few people use mathematics to any advanced degree and most of those who do concentrate on its applications in particular branches – statistics, economics, engineering. In spite of the centrality of mathematics to modern life, its practitioners exist in their own world and its communication with popular culture is partial and isolated. Some books about mathematics for the popular market assure the reader that they do not actually contain much in the way of equations. A mathematician in Cambridge is not as isolated as Archimedes only because there are more mathematicians in total and, especially, because mathematics now has institutional support, enabling concentrations of them to exist in academic departments and other research facilities.

In antiquity, mathematics had a rather similar reputation. Literary works which mentioned or made use of maths or astronomy tended to be inaccurate, such as the out-of-date Latin verse version of stellar movements and effects by the poet Aratus or the historian Polybius' lack of familiarity with astronomers' work on geography.[3] When Conon and Eratosthenes, both of whom relied on patronage from the Ptolemaic dynasty of kings in Greek-ruled Egypt, interacted with their patrons and the rest of the court, they presented their work as poetry and without actual mathematics. They did not expect an audience of non-mathematicians to understand either the methods or the conclusions of the work they had done.

Mathematics, particularly at its higher levels, is difficult. It is also more or less pointless to try doing it badly: a half-wrong equation is still wrong, while a moderate short story may still be enjoyable in some respects. According to a much later report, one of the Ptolemaic kings of Egypt is said to have asked Euclid if there was no shorter route to learning mathematics than reading the entirety of the *Elements*. He was told that there was no such shortcut, no 'royal road', in existence for mathematics.[4] This conveniently pithy anecdote expresses ancient attitudes to both royalty and mathematics – not least in its sly suggestion that kings want to avoid hard work. More seriously, it assimilates royalty to the ordinary person in their relationship to mathematics and insists upon the mountain peak reputation of the latter: a project for a prestigious yet marginal few.

The Rise of the Astronomers

As it was in Egypt, Babylonia and pre-classical Greece, mathematics is commonly used in all civilisations for the management and identification of resources, as well as in techniques for manipulating the physical world – engineering. The historian Polybius (c.200–118) could assert that historians and generals both need some understanding of geometry and astronomy as applied to tactics, camp-building and navigation of territory. The architect Vitruvius (active c.40–10 BC) also said that a knowledge of mathematics should be part of architecture. This is mathematics in the role that we would call 'applied', as vital to the design of cities, temples, villas and sewage systems.

Vitruvius, however, is also thinking of mathematical awareness – not the same as the capacity to carry out original mathematical proofs – as part of the proper education of the civilised man, that is as part of philosophy. Philosophers were not mathematicians, but the mathematical aspects of astronomy were thought to demonstrate, firstly, the order and rationality of the cosmos and, secondly, that true knowledge was possible. The very existence of mathematics as a successful field of study had implications for theology, philosophical epistemology and teleology. The idea, articulated by – among others – the astronomer Claudius Ptolemy and the physician Galen, was that the heavens demonstrated mathematical order to an unparalleled degree and that this reflected the planned or purposeful nature of the cosmos.

Astronomy played a crucial role in this perception. The positions and movements of the stars were guides to navigation and the social and agricultural year for a much older and wider world than the small circle of Greek mathematicians. Hesiod's seventh-century BC poem *Works and Days* reminds its hearers to bring in the grape harvest when Sirius is in the middle of the sky. The relationship of the heavens to the year, more precise than the seasonal changes, is the basis for every calendar.

According to popular anecdotes about him, the 'first Pre-Socratic', Thales of Miletus (*c*.624–546 BC), made a killing on the olive markets due to his ability to forecast astronomically influenced weather and also predicted an eclipse of the sun. Neither of these stories is likely to be true, but they highlight several aspects of early astronomy. Prediction was extremely important as a demonstration of knowledge, especially prediction of a dramatic event like a lunar or solar eclipse. A great part of the field of astronomy concerned the prediction of stellar movements on the basis of observational records and it was these predictions that established astronomy's authority – and that provided the questions which more geometrical and theoretical astronomy tried to answer. Secondly, it was commonly thought that stellar motion affected or at least correlated with meteorological patterns and events. That is a reasonable supposition. On a large scale like that of the seasons, the astronomical year and the climate do tend to agree. Ideas about the physical scale and structure of the outer cosmos supported this, since according to (non-atomist) theories the stars are very close to the world's atmosphere. Ideas like this provided the theoretical underpinning of later Greco-Roman astrology (see Chapter VI).

The most pervasively important role of astronomy, however, was its relationship to the calendar. In both classical Greece and Republican Rome time passed according to a variety of local civic calendars. In practice these worked by counting days in between each religious festival, historical anniversary, tax collection or political election. Formally, they were luni-solar calendars, in which a year consisted of 12 lunar or 'synodic' months (from new moon to new moon). A strictly lunar year is only 354 days, and so falls short of the solar year (just over 365 days). Over time the official calendars of all Greek city-states drifted backwards from the astronomical year and the arrival of the different seasons, with the consequence that a festival celebrating agricultural harvest no longer correlated with the actual time of harvest.

The problem could be solved by adding an extra thirteenth month about every three years. In Athens this was done by repeating one of the usual 12 months as and when necessary, so that (their equivalent to) June, for instance, happened twice. The aim was to keep the beginning of the year in line with the summer solstice, when the sun's position in the sky at rising and setting ceases to move northwards.

This process was carried out on a rather ad hoc basis and extra days were sometimes added by the relevant official if the usual time of a festival became inconvenient – the festival happened on the same date, but an extra day or two occurred before it. Different cities managed their calendars differently and the year did not begin at the same point everywhere in Greece.

The calendar became a matter of confusion and contention. In *The Clouds*, produced in 423 BC, the satirist Aristophanes refers to the contemporary Athenian festival calendar being out of step with the actual phases of the moon and hence with the divine calendar of the gods:

> And she [the moon] says she confers other benefits on you, but that you do not observe the days at all correctly, but confuse them up and down; so that she says the gods are constantly threatening her, when they are defrauded of their dinner, and depart home, not having met with the regular feast according to the number of the days. And then, when you ought to be sacrificing, you are inflicting tortures and litigating. And often, while we gods are observing a fast, when we mourn for Memnon or Sarpedon, you are pouring libations and laughing. (615–23)

Matters could have been simplified by switching to a solar year calendar of 365 days, like that of the Egyptians. This provides a much closer match to the astronomical year of just over 365.25 mean solar days, the period of the earth's orbit around the sun. Its periodic readjustment requires intercalary days rather than months. In antiquity, the number of days in the solar year was worked out as the number of days between one summer solstice and the next or between one spring equinox and the next.[5] Later astronomers mention the observation of the summer solstice on 27 June 432 BC by Meton and Euctemon of Athens. Meton was also said to have used this and other such observations in working out a more useful calendar, one based on a 19-year solar cycle.

Over 19 years, the difference between the actual number of days that has passed and 19 years of 12 lunar months is a total of seven extra months, which can then be inserted into the calendar at regular intervals. The advantage of using this 19-year period is that the extra months total a whole number rather than a fractional one, making the necessary additions much simpler. The system was not, however, utilised. Meton appears in Aristophanes' *The Clouds* – the same play in which the dramatist points out Athens' calendrical headaches – as another impractical dreamer making various Heath Robinson contraptions.

In Rome, the situation was similar until the middle of the first century BC. Its priests managed a luni-solar calendar of 12 months for a year of 355 days, and to make up the ten days annual shortfall inserted an extra month of 28 or 29 days after the 23rd day of February, while removing the five remaining days of February (which usually totaled 28). Over a four-year period this made the year an average of one day longer than the tropical solar year. But by the end of the Roman Republic this point had become moot. Inefficiency or corruption on the part of the priests responsible meant that the necessary year-by-year adjustments had not been made, and the civic calendar and the seasonal year were a good two months apart.

Julius Caesar took advice, and the advice he took was that of a Greek expert in the by now authoritative field of astronomy, Sosigenes of Alexandria. The Roman calendar became a solar one with 365 days, with a leap day to be inserted into February once in every four years. In addition, the calendar underwent a year of emergency surgery in 46 BC, with two additional months between November and December. The vernal equinox was thereby restored to March. The priests initially had trouble implementing Sosigenes' system (counting inclusively, they incorrectly added a leap day every three years) and further corrections had to be made in 8 BC. After this, however, the Julian calendar, an invention of Greek astronomical expertise and the trust given to this by the Roman elite, became established as the temporal framework of the Roman empire and subsequently of the western world.

Its chief defect, the extension of the actual tropical year by 0.0078 of a day, only becomes apparent to experience over several centuries. By the sixteenth century the equinox had moved again and was now occurring some 12 days earlier, on 11 March. This affected the celebration of the Easter festival, which was calculated in relation to the spring equinox and also to the Jewish festival of Passover, which uses a luni-solar calendar. Amongst the resulting reforms of the ecclesiastical calendar under Pope Gregory XIII in AD 1582 was a straightforward adjustment to the solar year: over each 400 years, three of the 100 leap days were left out.

The story of calendar confusion and reform in the ancient world highlights how time, like space, is not simply an objective aspect of the cosmos but also a social and political construct. Astronomical time is the result of the fundamental physics that control how two massive bodies relate to each other. The unique conditions of the earth's movements in space produce the rhythm of day and night, the seasonal weather of the Mediterranean, the precise patternings of the solstice, equinoxes and changing constellations of the night sky.

Human time is derived from this cosmic pattern, but reinscribes the complex astronomical periodicities into an organisational framework driven by our social needs. This was even more the case in antiquity than it is now. The Greeks and Romans lived in a world dependent for its agricultural produce on the natural rhythms of the seasons, without either freezers or greenhouses. They navigated by star and sun as well as by landmark, without compass or sextant or accurate clock. They had little artificial light and no light pollution, making the stars a far more visible presence and maximising awareness of the heavenly markers of time. The names of the constellations signified not the peculiar tales of a long-lost civilisation, but the socially validated stories of religion, literature, history and childhood.

Calendars construct time as a dimension of human society, breaking it down into regular amounts of years, months, days and hours that smooth out the physical irregularities of the universe into the politics of civic order. The control of time in this fashion is a matter of institutional and political power. In the fifth-century BC Greece that Meton lived in, each city-state (*polis*) had its own calendar and adjusted to astronomical time on – so to speak – its own timetable. By Caesar's lifetime (100–44 BC), the Roman empire dominated the lands around the Mediterranean basin, including the cities of the province of Greece. The Julian calendar erased traditional parochial Roman time-keeping, just as Caesar swept away the forms and institutions of the Roman Republic in favour of his own rule, and extended a universal calendar throughout its dominions as another aspect of Roman hegemony: a means of homogenising and ordering the empire according to imperial Roman values and needs. Our modern Gregorian calendar, like its predecessor, indicates a commonality of culture among its adherents. Those religions or states that use other systems assert a different identity, rooted in a different history of time.

Caesar's reforms show something else as well. Authority had shifted from religious authority and civic officialdom to the astronomer. Sosigenes had a Greek name and came from Alexandria in Greek-ruled Egypt, a centre of astronomical research from the third century BC onwards. Astronomical and mathematical expertise had created a new international technocracy, usurping the traditional prerogatives of an earlier kind of technical expert: the priests. This degree of cultural prestige shows that astronomy had come some way since Meton's early reforms.

In the modern world our calculations of physical time get ever more precise, but this has not removed the need to manage civic time. One problem is adjusting time at one location to time at another, since both seasons and daylight vary with geography. Crossing from one side of the United States to another, across a unified political entity and a continuous land mass, entails changing time zones three times, from Eastern to Central or Mountain, to Pacific time. These zones demarcate human social groupings and a structure of combined bureaucratic and technocratic decision making, since the lines mark an abrupt, arbitrary boundary between one temporal order and another. (If they were calculated on the scientific basis of longitudinal divisions around the earth, that is the time taken by the sun to cross an arc of the sky, they would be rather more evenly spaced than they are.)

Yet in spite of the overarching presence of the Gregorian calendar itself, local time still marks out geographical identity: east, west, centre. What time it is, or even what season you are in, is dependent on where you are. In this we can see that our constructions of time are still ultimately tied to the physical universe and the structure of the solar system, so that each point on the earth is expressed as much by time as by place: both are relative to the light and heat of the sun.

The nineteenth-century British Navy measured its longitude from the Greenwich meridian, the imaginary north–south line running through the Royal Observatory in that southern district of London, defining it as 0°. Britain's naval and political power transformed this cultural artefact into an international convention. Calculating longitude depended on accurate time measurement at sea, a problem which was not solved until 1773 (and remained difficult for some time after), so the moment when the sun crosses the Greenwich meridian is also a temporal reference point. Since the varying gravitational forces of the other bodies in the solar system cause the earth's velocity to vary as well, and because of its axial tilt, the path of the sun is not stable and modern Greenwich time is an average of its actual routes – hence Greenwich *Mean* Time. Greenwich time was used by nineteenth- and twentieth-century astronomers as the overarching temporal structure for the earth and solar system and was replaced only when atomic clocks provided a more accurate method of measuring time, in 1972.

Even atomic clocks, however, are averaged in order to provide modern 'Universal Time' and placed at mean sea level to correct for gravitational variation. Meanwhile Britain, which has kept Greenwich Mean Time as its winter calendar, changes this to British summer time for half the year with the addition of an hour and over the course of the twentieth century has experimented with several variations on this theme. Advocates of further reform are pitted against geographic constraints, as the effects would differ sharply from Scotland to Sussex.

In both antiquity and the contemporary world human societies impose their own somewhat arbitrary order, according to their needs and culture, on the effects caused by the earth's deterministic but incredibly complex path through space. The Julian calendar marks the first moment in Western history in which astronomy superseded other kinds of expertise in defining time (and place). It enabled, at first for the Roman empire and eventually on the worldwide scale of Universal Time, the establishment of temporal measurement on a global rather than only a local scale by establishing larger, consistent, mathematically based reference points; from which local timescales could diverge on a relatively organised basis, as time zones do today. Mathematical astronomy locates us precisely in the cosmos and on the planet, but it has also been integral to the organisation of society and our daily life.

Mapping the World

> Anaximander of Miletus, the pupil of Thales, was the first to depict the inhabited world on a chart [...] Now the ancients drew the inhabited earth as round, with Hellas in the middle, and Delphi in the middle of Hellas, since it holds the navel of the earth.[6]

The geographer Strabo (64 BC–AD 24) argued that the science of geography, literally 'writing about the earth', began, like much else in Greek culture, with the poetry of Homer (which is now thought to have been composed in its final form around the eighth century BC). The description of the earth embedded in Homer's poetry was said to have been given pictorial form by the Pre-Socratic Anaximander of Miletus in the sixth century BC, when he created the first world map. Ancient authors say that Anaximander's map was improved upon and superseded by the work of a man from the same city, Hecataeus of Miletus (*c*.550–476 BC), who produced the first works dedicated to geography in the following generation.⁷

Living in the trading city of Miletus on the Asia Minor coast, Anaximander might have known of similar depictions of the world in the older cultures around the Mediterranean. One of these survives. The Babylonian *imago mundi*, also from the sixth century BC, was first published in 1899. Reconstructed, it shows Assyria as a land mass, marked with several cities, surrounding the position of Babylon on the river Euphrates. Assyria itself is completely encircled by a river, beyond which are seven islands, described in an accompanying text. The totality is in the shape of a symmetrical seven-pointed star.⁸

Anaximander's world map has not survived, but we can reconstruct some aspects of it from other evidence about early Greek geography. The Greek conceptualisation also viewed the earth as surrounded by a great river, Ocean, as described in Homer, although there were no islands beyond it. Within the boundaries of Ocean was the 'inhabited world' or *oikoumene*, which consisted of the three land masses of Europe, Asia and Libya (north Africa). The Aegean sea was more or less at the centre, where it separated Europe from Libya with Asia to its east.⁹

The fragmentary passage of the geographer Agathemerus (active around AD 200), which I quoted above, says that the actual centre of the world was Delphi, the important religious centre and site of the most prestigious oracle of Apollo. This is probably a supposition of Agathemerus based on later maps, when the Delphi tradition was well-established. Anaximander may not have identified any specific point as the centre of the earth. But the general perspective of both his world-map and subsequent geographical writings was of Greece at the centre of things.

The focus of any society starts from itself and moves out, constructing the world in relation to itself and privileging certain aspects of it. This is a feature not only of the ancient world but even of modern maps, since projecting a three-dimensional globe onto a two-dimensional surface causes distortions in the relative sizes of parts of the earth. On Mercator maps, which use a method of projection devised by Gerardus Mercator in 1569, the size of areas nearer the poles are exaggerated in comparison to those nearer the equator, effectively shrinking Brazil by five times relative to the more northerly Alaska. Mercator maps can therefore be read as (accidentally) reflecting a geopolitical concept in which Europe and North America are emphasised at the expense of more southern countries. It is also conventional, in the United States and Europe, to put the northern hemisphere 'on top', although in space there is no up or down.[10]

Seven hundred years after Anaximander, the astronomer Claudius Ptolemy (AD c.90–168) described a considerably expanded and better known world, with longitudinal and latitudinal co-ordinates, measured in degrees and minutes, for 8,000 places. The maps of his *Geography* are lost, but can be reconstructed from the text. Instead of the Greenwich meridian, Ptolemy counted from the meridian that marked the extreme west of the known world. There were reports of islands in this location, situated in the Atlantic west of the Straits of Gibraltar. Greco-Roman writers identified these, about which they knew little and to which few if any people from the empire had travelled, with the legends of idyllic islands that appear in several European cultures: the so-called Fortunate Islands or Islands of the Blessed. The geographical entities themselves are in the right location to be the Canary Islands, Azores, or the archipelago of Madeira. The eastern world now reached to the Indian Ocean and the South China Sea; to the south Greco-Roman geography postulated a vast (and non-existent) southern landmass.

During the centuries that had intervened between Anaximander and Ptolemy, exploration and trade had filled in and refined much of this world picture. It had also considerably expanded it. The strange monsters and legendary peoples of Greek and other cultures were continually being pushed back further and further, to the mysterious margins of the world – including the Fortunate Islands. At the same time, geographers and explorers tried to identify these places with actual, mappable locations. The Hyperboreans, the people said to live beyond the north wind, were variously placed around the Black Sea area or – by later authors – in the western Alps, and eventually somewhere north-east of Britain, perhaps in the southern Baltic.

In Homer, Ethiopia was a utopian society of plenty and equality, where the gods went to rest when exhausted by mortal and family conflicts. In Herodotus' fifth-century BC *History*, the Ethiopians are the Nubians, living south of the first Nile cataract. His sources in Egypt told him of their vast wealth, huge elephants, painted bodies and unusual tallness and longevity. Legend and geography meshed to produce the Greek genre of ethnography, the ancestor of anthropology. Ethnographies like those of Herodotus described other societies – from a Greek perspective, but ostensibly as an objective witness – in terms of environment, social practices, physical characteristics and culture.

The maps of the *Geography* were in a format similar to the atlas, while Ptolemy's city list is very like a modern gazetteer.[11] Greek geography seems to have adopted a realism-based approach from an early stage, so that regions were portrayed with positions and sizes as correct as current knowledge allowed, rather than in more symbolic or schematic depictions.[12] On the other hand, some maps or geographical works had practical rather than accurately descriptive uses. The Peutinger Table is a medieval copy of a map of the Roman empire, based at some remove upon a lost fourth-century AD map. It comprises a rectangular strip nearly seven metres wide but only 34 cm high, with the result that north–south distances are drastically compressed relative to those on an east–west axis, and the geographical landscape is consequently distorted. But this did not matter, because its purpose was to indicate distances between places and give some illustration of their direction and course, rather than provide a two-dimensional rendition of the actual landscape. In a similar vein handbooks were available that listed places by their distance or travel time from other locations, and astronomer-geographers often gave the longitude of a place in terms of its time difference from Alexandria.

An obvious comparison are the charts of driving time for particular distances to be found in modern atlases. Taken further, the realisation that the useful information in a map need not be rendered in photographic style leads to schematics like that of the London Underground created by Harry Beck in 1933. He based his design on circuit diagrams, so that the Tube map shows commuters what stations and lines they need rather than a literal representation of the Underground's physical tunnels.

Works of this type did not need to rely upon the findings of astronomical geography or upon the knowledge accumulated by ethnographic and exploratory geography in the descriptive tradition of Anaximander. Local knowledge, trading knowledge and the infrastructure and organisation of empire were sufficient to produce such gazetteers and route maps.

Like gazetteers, astronomical geography sought to orientate and locate humans through a process of quantification, but it worked upon a very different scale. Developing with the rest of astronomy and mathematics as part of naturalistic philosophy and Greco-Roman intellectual culture, it aimed at understanding rather than use, although celestial mapping did indeed have uses in terrestrial geography.[13]

The notion, derived from the divine inspiration of the poets, that Ocean encircled the inhabited world, remained strongly embedded in Greek thought, but the sphericity of the earth was recognised as early as the fourth century BC and possibly before. (Anaximander, incidentally, had thought it was a cylinder with a breadth three times that of its depth; what we see as our world is one of the two flat surfaces.) Strabo pointed out that as a ship sails away from an observer over the horizon, its hull disappears first and Aristotle too had given persuasive empirical arguments in favour of sphericity. One of these is that the traveller sees different constellations as he moves north or south, which would not be the case if the earth were flat. And in Aristotle's most visually conclusive example, the earth's shadow is always a convex curve when it crosses the moon during an lunar eclipse.[14]

Mapping the Cosmos

As we saw at the beginning of this chapter, early Greek astronomy – as opposed to cosmology – developed around the fifth century BC partly as an interest in establishing the temporal patterns of the earth in relation to the heavens. In Athens, Meton and Euctemon used a simple astronomical instrument, the gnomon, to identify the summer solstice of 432 BC.[15] The gnomon is essentially a stick. Stuck vertically into the ground, it casts a shadow and, modified further, becomes a sundial. Watching the edge of the shadow reveals a lot about the motion of the sun throughout the day and the year. (This works better in Greece than Iceland or Manchester.) Meton and Euctemon could see when the sun ceased to move northwards.

To this geometrical picture of the sun's motion relative to the earth (or vice versa) the very early astronomers had an aid to quantification in the shape of the *parapegma*, to keep track of the days between solstices. A *parapegma* is a flat stone or wooden tablet with holes into which pegs can be inserted. The holes can be labelled or marked to indicate a particular point in time such as the days of the month and the peg moved as each day arrives, functioning as an almanac. The Geminus *parapegma* dates from the early second century BC or the first century AD and compiles data from earlier ones, including one by the solstice-observing Euctemon and another by the mathematician Eudoxus (*c*.408–347 BC). Part of the Geminus *parapegma* for the month in which the sun moves through the zodiac sky-division Virgo is as follows:

> On the 10th day, according to Euctemon, the Vintager appears, Arcturus rises, and the Bird sets at dawn; a storm at sea; south wind. According to Eudoxus, rain, thunder; a great wind blows.[16]

Many of the Greek constellations are from older Babylonian identifications, as is the division of the sun's path through the ecliptic into the 12 signs of the Babylonian zodiac (for astrology, see Chapter VI). The earliest Greek astronomers, lacking records of their own, relied on Babylonian knowledge of the movements and positions of stars, which a specialised group of priests had observed for centuries so that they could identify days of good and bad omen with certainty. By the Hellenistic period of the third to first centuries BC, precise observations and mathematical quantification were integral to astronomy and formed the basis for its explanations of the universe's physical and mechanical workings. Eudoxus' hypothetical model of 27 spheres, all centred on a stationary earth, was revised by other thinkers but by the second century BC had been replaced by theories that used the mathematically equivalent solutions of eccentric and epicyclic orbits for the planets, sun and moon. As improved by Ptolemy, this system dominated until the Copernican revolution. The heliocentric hypothesis worked out by Aristarchus of Samos (c.310–230 BC) in a lost treatise was rejected by most Greek authors due to its lack of empirical confirmation (stellar parallax was not visible to the naked eye) and its conflict with the standard, Aristotelian, theories of physics.

Astronomers also attempted to quantify the universe and map the cosmos by estimating, in the words of another of Aristarchus' treatises, *The Sizes and Distances of the Sun and Moon*. (His results were out by some margin, but calculating that the sun was much larger than the earth may have been a factor in him proposing a heliocentric model of the universe instead of the more usual geocentric one.) Half a century later Archimedes published his work on very large numbers and their notation in *The Sand-Reckoner*, which asked how many grains of sand could fit into the universe. And several attempts were made to work out the size of the world.

The best known of these is that by Eratosthenes, from Cyrene in Libya, the polymath poet, mathematician, literary scholar, philosopher and geographer working in third century BC Alexandria during the rule of the

Greek-speaking, Greek-descended Ptolemies. Observations or reports told him that in Syene (modern Aswan) in south Egypt the sun at noon cast no shadow on the day of the summer solstice, meaning that it was directly overhead. On the same solstice day further north, in Eratosthenes' Alexandria, the sun at noon (its zenith distance) did cast a shadow for an upright gnomon, which Eratosthenes measured at 1/50 of a circle or 7.2°. Eratosthenes knew the distance between Syene and Alexandria: in local measurement, 5,000 stades.

He then used relatively simple geometry to show that the circumference of the earth must be 50 times the distance between Syene and Alexandria, taking them to be on the same north–south meridian. The figure is 250,000 stades. Unfortunately, the ancient unit of measurement called a *stade* (derived from the length of an athletics stadium, and thus an ancestor of the colloquial modern tendency to measure things in ball parks and football grounds) varied considerably according to location. We are not sure which one Eratosthenes was using. Of the available options, one gives his result as 24,633 miles and the most inaccurate would be 27,967. Today, the equatorial circumference is 24,902 miles.[17]

It seems likely that Eratosthenes thought this an approximate estimate because one source says he used the figure of 252,000, and the obvious reason for doing this was to make it divisible by 60 or 360, in accordance with the multiples-of-12 based astronomy that the Greeks had inherited from the Babylonians.[18] Of the other estimates available in antiquity, we know of two prior to Eratosthenes that are considerably larger: 400,000 and 300,000. We do not know by what method these calculations were made. Later, the philosopher and natural scientist Posidonius of Rhodes (135–51 BC), using calculations based on the relative positions of the star Canopus at Rhodes and Alexandria, arrived at a similar result to Eratosthenes': 240,000 stades. But another source says that Posidonius worked out the circumference to be a much smaller 180,000 stades, probably on the basis of a lesser figure for the distance between Rhodes and Alexandria. Claudius Ptolemy preferred the 180,000 figure, and this became the commonly cited distance for the earth's circumference in later antiquity and the Middle Ages. By then yet more estimates were in circulation on the basis of calculations by Arab astronomers.

Empire by Numbers

The cosmic quantification of the universe, carried out by the mathematically adept, was also an increasingly visible feature of life on earth. The two projects could intersect and inter-relate, as they did in Eratosthenes' and Posidonius' calculations of the earth's circumference – calculations which utilised known distances between two geographical reference points as well as the date of the summer solstice and basic trigonometry. For the Roman empire, mapping, counting, calculating and engineering all proved vital tools in the acquisition of information and the maintenance of control; tools that required an expanded role for the technical expert and the more common forms of mathematics.

Greek city-states and the Roman Republic had kept and sometimes published (as inscriptions) their government accounts, but by 30 BC the size of the Roman empire, as Octavian Augustus took control, required its own small army of numerate officials and their secretaries. They managed the empire-wide census, the tax rolls, the land surveys and the army's accounts. Such financial officials were stratified and organised. Public accountants (*numerarius*, *tabularius*), with the assistance of secretaries (*scriba*), worked for the high-ranking men 'from the accounts office' (*a rationibus*). Large private households and businesses had similar staffs. Some of the lower status secretaries and accountants were slaves, taught special skills so that they could fulfil a necessary but specialised function. Many were freedmen: upwardly mobile agents of the state apparatus. Some gained considerable power through such routes, including access to the emperor. Traditional social networks of influence found themselves facing unfamiliar competition and many aristocrats resented the power of the imperial staff. Unlike an older kind of 'new man' (*novus homo*) such upstarts did not always adequately conceal their origins or take pains to integrate themselves into the old order.

A higher-status mathematics, largely in the form of geometry, was also part of the expertise of architects and engineers, as Vitruvius makes clear at the start of *On Architecture*:

> Geometry, also, is of much assistance in architecture, and in particular it teaches us the use of the rule and compasses, by which especially we acquire readiness in making plans for buildings in their grounds, and rightly apply the square, the level and the plummet. By means of optics, again, the light in buildings can be drawn from fixed quarters of the sky. It is true that it is by arithmetic that the total cost of buildings is calculated and measurements are computed, but difficult questions involving symmetry are solved by means of geometrical theories and methods.[19]

Vitruvius worked as both an architect and a military engineer during the first century BC, dying at least 15 years into Augustus' reign. *On Architecture* covers 'how to build things' in many more fields than the modern reach of the term, including catapults and sundials.[20] Geometry and arithmetic offered Roman power two instruments of control over the landscape, from the largely hidden infrastructure of its water supply and sewage systems to the army fortifications and new grid-based cities of its empire.

A classic illustration of the former is provided by the career of Sextus Julius Frontinus, a top-ranking official and engineering or infrastructure specialist. In AD 97 he was appointed water supervisor of Rome itself. Having surveyed Rome's nine aqueducts and calculated their capacity, he compared this to the city's output of water and found a sizeable discrepancy.[21] His conclusions were right – private individuals were siphoning off water and bribing water officials – although his mathematical method was wrong: water output cannot be calculated from the diameter of pipes. The significance is that Frontinus perceived mathematics as the way to understand the situation and to demonstrate his credentials as a good administrator. By quantifying water, he could quantify corruption.

He solved the situation with an appeal to mathematical order, by standardising the diameter of Rome's water pipes.

The engineers and surveyors of the Roman armies and provincial administrations had a similar approach to managing the empire: measure, calculate, standardise. Using instruments like the sighting-tube of the *dioptra* – a design which earlier mathematician-engineers like Hero of Alexandria had worked upon and which was also used in astronomy – imperial surveyors, *mensores*, plotted the landscape into a measured grid of *centuria* (each two-thirds of an acre). As Serafina Cuomo puts it: 'the centuriated territory became in effect a geometrical landscape'.[22]

Once the emperor and his agents knew how much of what kind of land was where and belonged to whom, they could assess, redistribute and take away as they liked. This was not a new means of control – societies like Egypt and Babylonia had used geometrical formulae and arithmetical calculations to manage land and population in much the same way millennia earlier – but the Roman administration's efforts were on a different scale.

One effect was to standardise the Roman world. Within the borders of empire there was a mathematical and administrative unity of form. Since military camps, new cities, villa or temple designs and infrastructural elements like aqueducts were largely built to a relatively small set of templates, there were also elements of the Roman world that looked visually similar from place to place. (In the twenty-first century, one result of the western world's classical tradition is the pervasive presence of Greco-Roman style columnar architecture, usually on important public buildings and aristocratic efforts to impress.)

Vitruvius' and Frontinus' mathematical architecture, combined with the organisational manpower of the army, enabled technically difficult or massive projects. An inscription of AD 152 describes the difficulties in driving a water-tunnel straight through a mountain. Dug from both ends, the tunnel was not meeting in the middle and an engineer from the Third Legion, Nonius Datus, had to be sent. He complained that such problems were always blamed on engineers, but he resolved this case with a connecting section of tunnel.

Meanwhile, in northern Britain, the remains of the great borderland wall built across it from coast to coast under the emperor Hadrian still survive. The wall was ostensibly there to hold back the Picts of Scotland. It was also a visual declaration of Roman power, endurance and technical expertise. The landscape was not only geometrised, it was engineered and turned into a form of political display.

The same trends towards quantification and technical knowledge can be seen in civic and private life. Conversion tables were circulated as a solution for coping with different systems of units, whether of measurement or of money. Such unplanned inconveniences of technical history or economics survive today, embedded as a cultural preference for Fahrenheit over Celsius or the United States's dislike of the 24-hour clock. They cause confusion to holidaymakers and occasional problems for NASA.[23] The solution is still often the conversion table, which has been joined by computer programs and smartphone applications. In classical antiquity portable bronze abaci seem to have been widely used and sundials, some technically sophisticated, are found in both private and civic spaces in the remains of Roman cities. This astronomical-mathematical device acted as a local standardisation of time in markets and public squares.

Since it began to evolve as a specialist discipline in the sixth and fifth centuries BC, mathematics had transformed itself by having a considerable effect on the world around it. In particular it became integral to the physical and social empire established by Rome. The mathematics instantiated in land measurement, calendar reform and the civilian and military infrastructure built by the engineers lent a shared structural homogeneity to the empire's physical and institutional landscape: from the centrally heated villas that marked the wealth of the empire's elite to the roads that enabled travel, trade and communication across its (measured) length and breadth. Distance and time became co-ordinates of the same social polity. Quantification, vital to administration, also became a personal way of enumerating and comparing achievement – of literally measuring success.[24] Elements of mathematics, including the more theoretical aspects of astronomy, remained the preserve of the privileged: a kind of intellectual kite mark. But mathematics as a whole, from astronomical geography to mechanics, was a technical expertise of much wider value to society and the state; and a vital component of the increasingly global reach of Rome.

The Rise of the Machine-Makers

Astronomy was not the only means by which apparently abstruse questions in mathematics and natural philosophy had ramifications for society and culture. The Greek classical period had seen the emergence of a discipline known as 'mechanics'. It combined geometry, philosophical theories of physics and older technological expertise to become a science of machines and of the underlying mathematical and physical principles that explained why and how such devices worked.

Mêchanikê is a term derived from *mêchanê*, the Greek for 'machine, device'. A famous example of a mechanical device in Greek culture was the hollow wooden horse employed in the Trojan war to smuggle soldiers into the besieged city; a story alluded to in the eighth-century BC *Odyssey* and in literature from classical Greece, although it is best known from later Roman retellings. Its invention was linked to Odysseus, the Greek hero characterised by guile to the extent that one of Homer's poetic epithets for him was *polymechanikos*, 'the man of many tricks'. This gets at what for us is an ambiguity in the general meaning of the word *mêchanê*: it can mean either a stratagem or a physical machine and the Greek encompasses both senses without presenting them as alternatives.[25]

The Trojan horse is a good example of this ambiguity because it was both a cunning plan and a physical, functional instantiation of that plan, which concealed its purpose and produced a non-obvious end result. This association of machines with surprise and cunning – not to mention warfare – remained a feature of mechanics as a discipline.

The first person in extant texts to use the term mechanics is Aristotle (384–322 BC). By then the subject was already well-defined enough for him to describe it as the subordinate discipline to solid geometry, in that the former deals with particulars and the latter with universals (*Posterior Analytics*, 1.13 (78b38)). Aristotle might also have been the author of the first known treatise on the subject, *Mechanical Problems*, although it is more likely to have been the work of someone from his school of thought, composed perhaps in the early third century BC.

Thus by the time that mechanics is visible to modern eyes in extant writings it is already a defined field of investigation. We know little about the when, why and by whom of its pre-Aristotle antecedents. What is clear is that it was viewed as being a close relative of mathematics and, in particular, solid geometry; and that it was also largely concerned with how mathematics and natural philosophy applied to the physical workings of machines and mechanical elements; especially machines that produced movement, like the lever, axle and siphon. Later mechanists built on this by producing such machines as water organs, siege engines, stage machinery for theatrical performance and self-moving statues.

In its broad outline this double aspect of mechanics remained constant throughout its subsequent history in antiquity, so that Pappus of Alexandria, writing a compendium of mathematics in about AD 340, described mechanics as having both a practical dimension and a theoretical one, with the theoretical aspect being largely mathematical in nature.

Those individuals significant to the history of Greek mechanics sometimes specialised more in mathematics, sometimes in the actual construction of machines. Archimedes (*c*.287–212 BC), for instance, can be included in such a list not because he self-identified as a mechanist or even because later and rather unreliable tradition cites him as the creator of several startling feats of engineering, but because some of his mathematical work concerned principles of physics and mechanics.[26] Of his surviving treatises, *On Floating Bodies* discusses the principle that the buoyant force acting on an object in a fluid is equal to the weight of the fluid displaced – the principle more famously recalled in the story of Archimedes crying 'eureka!' in his bath. *On the Equilibrium of Planes* concerns problems of balance that machines put into practice. Lost treatises by him seem to have involved the mathematical principles behind machines that lifted weights, a category picked out by a later author as a particularly useful branch of mechanics.

One person sometimes identified as a founder of mechanics, or at least as crucial to its early evolution, is the statesman, philosopher and mathematician Archytas of Tarentum (c.435/410–360/350 BC).[27] Archytas held the very high rank of *strategos* ('general') in his home city Tarentum, a Greek colony in southern Italy, for ten years, at a time when that city's power and influence in the western Greek world was at its height. In addition to being one of the leading geometers of the time, Archytas was also well-known as a theorist on ratio and proportion in music, on ethics and civics, on at least some aspects of cosmology and as a Pythagorean philosopher.[28] An approximate contemporary of Plato (427–347 BC), his views were discussed by Aristotle and his reputation was sufficient to ensure an afterlife in Roman prose and poetry, among both 'scientific' authors like Vitruvius and Pliny and in works by poets and antiquarians. Fragments of his writings survive in these and other authors throughout the Greco-Roman era. But as a result of this long process of transmission we can only be sure of a few surviving textual fragments as being to some degree representative of his original words and views.

Mechanics emerged as a theoretical discipline contemporaneously with Archytas' career, in the middle and later fourth century BC. But there is not much direct evidence available for assessing whether or not he can usefully be described as the 'founder' of mechanics, as some have argued. Later sources say that he was the maker of a flying wooden dove and a toy. This is suggestive as far as it goes, but not in itself demonstrative of the kinds of interactions between geometry, physics and materials that were utilised by mechanists in subsequent, better-known generations.

We can be more confident in a vaguer claim, that fourth-century BC developments in mathematics – not only in geometry itself but also in what Archytas called 'logistic', the mathematics of ratio and numerical proportionality – became a definitive part of the emerging expertise of mechanics and its utilisation of geometrical solutions.

Cubes, Catapults, Computers

This can be seen in the history of a mathematical problem known as duplicating or doubling the cube. It is a problem in geometry of the kind that seems simple when posed but defies common sense attempts to answer it. If one has a cuboid object of a particular size, how does one work out what could be the measurements of a cube with double that original volume? For example, suppose that the original cube measures $3 \times 3 \times 3$, in whatever units you like, giving it a volume of 27 of those units. A cube of twice that size has a volume of $2 \times 27 = 54$. But the lengths of the sides of the cube have not doubled. If they had, that would mean a cube of size 6^3 and would equal 216 units – not twice but eight times the original size. Alternatively, adding a second cube of the same size to the original produces the right volume ($6 \times 3 \times 3 = 54$) but now you have a rectangular object, not a cube.

'Doubling the cube' is therefore part of a bigger set of problems: how to change the size of three-dimensional objects without changing their shape: scaling something up or down when form mattered for function but size was allowed to vary. For example, statues were often carved from smaller models: creating the need to make something bigger without changing the proportions of its parts relative to each other. In architecture the building of temples produced similar difficulties in scaling, and in fact the origin story for 'duplicating the cube' is a legend in which Apollo ordered a cube-shaped altar of his to be doubled in size without changing its shape.

Among the mathematical solutions that survive from antiquity, Archytas' is the most imaginative and the most difficult to visualise, involving as it does a set of imaginary operations in three dimensions. In brief, Archytas constructed four triangles by means of an intersecting point of two rotating plane figures, in different planes, that trace out a line on the surface of a semi-cylinder. For each triangle thus constructed, the length of the original cube's size forms one side and double that length another side. With this construction he could show that the triangle sides were serially proportionate to each other and which side would create the doubled cube.

But if it was the first answer, it was certainly not the last. Over several centuries Greco-Roman mathematicians produced some 11 different solutions to the problem of doubling the cube. Each answer had its own strengths and weaknesses and which was best depended partly on the reason for asking the question. Archytas' version, a geometrical tour de force, was difficult and abstract enough that it might as well have been labelled 'for really good mathematicians only'. The author of a later and very different solution, the polymath Eratosthenes (276–194 BC), criticised it for not being easy to put into practice and, in a probably genuine letter to the Greek king Ptolemy II of Egypt, emphasised the theoretical aspect of all his predecessors' work on the problem of doubling the cube:

> they have all written in the form of a geometrical demonstration and they cannot build what they describe [...] But I have contrived an easy approach, by means of an instrument [...] With this discovery we will be able in general to transform any given solid which is bounded by parallelograms into a cube [...] so as [to make] both altars and temples.[29]

He describes such answers as *dusmechanikos* – 'not at all mechanical'. Eratosthenes' own alternative was, in deliberate contrast, a mechanical solution that used a wooden frame containing three parallelograms as a sort of three-dimensional slide rule: an input and output machine that did the calculation for you if you moved the parts according to a simple, easily learnt set of operations.

And, as he explained, the usefulness of such a method is that it is not limited to making cubes twice as big: it can scale any solid object up or down while keeping its structural parts in their original ratios to each other. Eratosthenes cited religious architecture as his area of applicability, but one of the chief uses of a readily applicable method for scaling a particular object up and down turned out to be war artillery used in sieges – specifically, machines that catapulted stones or other missiles by means of a torsion mechanism.

Torsion artillery, which was not yet developed when Archytas was devising his proof for doubling the cube, depends on a spring made of sinew or hair. Its compressed kinetic power enables catapults to throw stones and other shot, including sharp objects, with more power than even the largest practicable tension artillery. One expert on catapults argues that a crucial step was the discovery, sometime between about 320 and 310 BC, of this 'scaling law' as applied to torsion catapults. Experimental versions of a catapult, for instance one that had worked well in battle, could now be made larger or smaller without losing the balance between their structural elements that had made them good in the first place.[30]

So by the first century BC solutions to doubling the cube had mutated from Archytas' brain-twisting proof of concept, the answer to a largely theoretical puzzle, into a vitally important element of artillery design and construction among the political states of the period and their not infrequent wars. The case of the cube roots and the catapult is thus a leading example of how mathematics enabled mechanics, because mathematics provided a reliable set of truths about the way in which quantifiable objects interacted with each other; truths which were not limited to a particular instantiation of numbers but described the relationships between numbers and shapes. Mechanics relied on mathematics both for epistemological validity, equating to intellectual status, and for practical utility.

But the relationship was not one-way. In some cases machines could be used as instruments with which to do mathematical calculation. This aspect of the relationship between mathematics and mechanics was often a source of cultural anxiety, especially in the writings of certain intellectual non-mathematicians, like Plutarch. It evolved into the trope that Plato had disapproved of Archytas for doing this kind of thing, on the probably anachronistic presumption that Archytas had been some sort of mechanist himself.[31] Perhaps it was problematic because it implied that mathematics could use – even need – mechanics. This was a threat to the intellectual status accorded, in much Greco-Roman tradition, to abstract thought as pursued by those who did not have to work with their hands. It suggested that mathematics might become accessible to people with little or no mathematical knowledge of their own.

Eratosthenes' slide rule was one such machine, but they were most common in the field of mathematical astronomy. Astrolabes, invented around the second century BC, enabled a user to measure the altitude of celestial objects. A more complex kind of instrumentation was the armillary sphere, which modelled the cosmos as a small sphere (the earth) surrounded by four or six metal rings that could be moved relative to each other and the earth-sphere: one represented the meridian, another latitude, another the ecliptic, another the zodiac and so on. The stand on which the entire thing rested could be set as the horizon, giving the right angle of obliquity according to the observer's earthly latitude. The first extant ancient author to mention them is Geminus (AD c.10–60), but they were probably available considerably earlier than that. Some armillary spheres served principally as models, to demonstrate the arrangement and order of the cosmos or to work out relative positioning; larger and more precise ones could also be used for observational purposes. Sights on an inner ring enabled the user to select a particular celestial object and thus to determine its ecliptic co-ordinates.[32]

But the most startling example of ancient mathematical machinery is a device recovered in 1901 from the undersea wreckage, off the Greek island of Antikythera, of a Roman trading ship from the first century BC.

The so-called Antikythera device was probably made towards the end of the second or beginning of the first century BC, perhaps in Rhodes. It has 30 bronze gears and has lost at least five more to damage and corrosion in the seawater – its original wooden case being insufficient protection for 2,000 years underwater. The front displays two dials, one within the other, marked respectively with the divisions of the 365-day Egyptian year and the 12 divisions that constitute the Greco-Babylonian zodiac. The hands for these dials have not survived, but there were probably three: showing the date, position of the sun and position of the moon, relative to the calendrical divisions on the dial. The front of the instrument also included another mechanism, this one with a spherical model of the moon that showed its phase for any particular date, and a *parapegma* with which to mark the risings and settings of specific stars. On the back were 3 more dials: one displaying the 19-year Metonic solar cycle and the 76-year solar cycle worked out by Callippus. The other two featured eclipse cycles.

32 bronze gears within the mechanism, once it was in motion, turned the dials relative to each other, allowing the movements of the celestial bodies to be read off as they revolved. The power came from turning a small hand crank (not extant) that connected to the gears.

So, to take one aspect of this, the sun marker and the moon marker were driven by two composite gears, with the axis of the moon marker threaded through that of the sun's.[33] The sun gear's 64 bronze teeth meshed with the 38-tooth gear that was paired with a 48-tooth gear, so that these latter two moved in tandem. The 48-tooth gear meshed, in turn, with a 24-tooth gear paired with a 127-tooth one, and that 127-tooth gear meshed with the 32-tooth gear of the moon marker.

In any geared device, the ratio of their turn speeds – as the turn of the gear moves through the teeth of its counterpart – is determined by the distance from the centre of the gear to the point of contact. For example in a two-gear device in which one gear has twice the diameter of the other, the ratio of their speeds would be 2:1. In the elaborate sun and moon system of the Antikythera device, which has three two-gear devices in a chain, the ratio of the respective starting and finishing speeds is therefore (remember the teeth are all the same size): $64/38 \times 48/24 \times 127/32 = 254/19 = 13.2684211$.

This is almost exactly the actual astronomical ratio of the sun and moon's apparent speeds around the zodiac (the moon, of course, is the speedier one). This level of accuracy must have been largely achieved through astronomical records, repeated and refined until the relative speeds were known so precisely.

The Antikythera mechanism is the most complex machine known in antiquity; nothing comparable has survived in the historical record until a

Byzantine example of the fifth century AD. It may have been still more elaborate: it is probable there was gearing to show the positions of Mars and Venus and perhaps even the positions of the other planets as well, although reconstructions differ on how this would have been done and the resultant level of accuracy.[34]

The mechanism has been described as a computer, but not in the sense in which the word is used in the early twenty-first century, because when we say 'computer' we are referring to a digital and universal Turing machine. The Antikythera mechanism is an analog computer, not a digital one; and it is not Turing-complete.

An analog computer is a computing device that uses changeable physical states to model the bits of information in whatever problem it is trying to solve. A slide rule is an analog computer of a mechanical kind in which marks on the sliding central strip are aligned with a mark on a fixed strip and relevant marks elsewhere are then read off. The slide rule is therefore a model of the bits of the world relevant to the questions being asked, with those bits being represented by the physical organisation of matter. An analog computer might work on an electrical or hydraulic basis as well as on a mechanical one: the important thing is that its physical states are capable of continuous change, as when I move the sliding central strip of the slide rule along a continuous material line.

In an analog computer each computation is therefore a unique physical event. This means that if I repeat the same calculation, the computation will be almost the same physical event as before, but not quite – because it is impossible to literally reproduce a physical event twice without so much as an atom having changed. This inevitable, if minute, irreproducibility is called analog noise.

A digital computer, like all modern computers, does not suffer from analog noise because it represents change symbolically. In a digital machine you do not move continuously from three to four along a line or an electrical potential or as a change in water level; whatever you are using to represent three flips from one state to the other: three OR not-three, on OR off. One of the earliest digital computers was the Colossus machine developed by

British code-breakers in World War II, which used thermionic valves (vacuum tubes) as automatic electronic switches (current flowed between anode and cathode only if a third wire was at a certain potential). The switch was either on or off. Here the problem of analog noise disappears, because the physical state of the valve is the physical mechanism enabling the calculation, not the calculation instantiated in a physical state.

But even a Colossus machine was not a computer in the modern sense, in spite of being digital. It was designed to do the kind of logical operations relevant to World War II cryptoanalysis: it was not a general computational machine in which the on/off values could symbolise any possible input and therefore carry out any computation. In an influential paper of 1936 the British mathematician Alan Turing, whose ideas had influenced Colossus and the Bletchley code-breaking machines, described a device that could be programmed to digitally imitate the steps in any other computational device (real or imagined) and hence run any possible computation:

> It is possible to invent a single machine which can be used to compute any computable sequence. If this machine U is supplied with a tape on the beginning of which is written the S.D ['Standard description', which means the encoded description of any computing machine M] of some computing machine M, then U will compute the same sequence as M.[35]

A computer in the twenty-first-century sense is a Turing machine: it can simulate any kind of digital or analog computer by means of the program it is running. This makes it a universal computing machine, because the software is not limited to the hardware.

The Antikythera mechanism is thus much more like a very complicated slide rule than a Mac or PC or a smartphone, which does not make the achievement less remarkable. It was designed to perform a limited but complex set of calculations concerning astronomical bodies, for which it utilised the continuous change in physical state available to moving, interlocking gear wheels. If it had not been for the chance of the shipwreck, its discovery and the fortunate survival of much of the mechanism, we would have no evidence that Greco-Roman technology was capable of such things. It is both fascinating in itself and a useful reminder of how much of ancient science and technology we might underestimate or misrepresent, simply because of the passing of time.

In other fields, too, mechanics was capable of demonstrating the truth of mathematics in a way that could make sense to non-mathematicians. In antiquity, mechanics was less about empirically confirming a hypothesis than demonstrating its reality through applicability. Such mechanical or physical 'proofs' also have an advantage over a mathematical or logical one: they are much more accessible and often more persuasive.

Mechanics *dramatised* the power of mathematics, science and especially mechanics itself to explain and manipulate the world. It did so with actual devices, often one-off pieces for display or conversation like the water organ of Ctesibius; or with the ideas, plans and descriptions that comprise significant parts of Hero's written oeuvre, such as his spy-mirror system; or in the biographical tradition about Archimedes, with its light-ray weaponisation of mathematical optics (this chapter, n. 26).

But this dramatisation of the power of science was not limited to the imagined or fictional, or even to the unusual. When Archimedes said, 'Give me a place to stand and with a lever I will move the whole world', this was not a claim about what he could literally do.[36] Archimedes could not move the earth. For that matter, neither can we. Instead it was a deliberately extreme claim about the power and truth of mathematical physics, mechanised in three-dimensional form (the lever) and specific to a precise, quantified place in a mapped universe (the right place to stand). The *physics,* he is saying, is true even to this mind-boggling degree. Its reach is godlike, universal and real – and accessible only by the mathematician.

In July 1945, physicists stood in the New Mexican desert and watched a nuclear test bomb explode. It was the same design as the bomb dropped over Nagasaki, Japan, later that year. The leader of the research team, J. Robert Oppenheimer, said that at the time he thought of a line from the Bhagavad-Gita: 'I am become death, destroyer of worlds.'[37] It was not precisely a lever, but there is no more forceful demonstration of the destructive as well as the creative power of mathematics, as applied to an understanding of physics, than the mushroom cloud.

Formulas to Live by

Mechanics was particularly concerned with machines of motion and change, such as the lever, axle, balance and siphon. In Pappus' fourth-century AD list of what comprised mechanics he groups machines that do 'surprising' things like lift great weights (pulleys, cranes) or move themselves (automata); machines that do useful things (artillery, water screws); and, thirdly, models of the heavens, like astrolabes and the Antikythera mechanism. Such machines expanded nature's repertoire, sometimes to the extent of apparently working against it, as when a siphon or the Archimedean water screw moved water from lower to higher places so that the water literally ran counter to everyday observation as well as against Aristotelian physics (in which water moves innately to its natural place above earth and below air, and can only be dislodged by force).

In a way, then, mechanics used the natural laws of physics (remember that in Greek *physis* translates to something like 'nature') to accomplish aims that are apparently *contra physin* – against nature. This is not to say they were thereby unnatural. (Like us, Greeks and Romans had a tendency to view the natural state of things as a valued norm, and divergences from whatever this 'natural state' was construed to be, as the work of a degenerate, decadent culture, divorced from nature and overly artificial.)[38]

That persuading water to go uphill, or using a lever to shift more weight than seems reasonable, is not really 'against nature' is implicit in a guiding assumption of ancient mechanics: that its principles were to be found in the natural world, and vice versa. This is a statement not only about such things as the reality of bubbles of void in matter and how these can be used in technology, but also about mathematics being as real in the material and social world as it was as an abstracted set of rules and interactions between numbers, lines or purely symbolic items.

One of the reasons for Archytas of Tarentum being a famous mathematician, in addition to duplicating the cube, was his work on music ratios. The realisation that the pitch of a note is proportionately dependent on the length of the string and therefore that intervals between the sounds that musicians produced could be expressed mathematically – the science of harmonics – was credited to Pythagoras and his followers, and Archytas was some kind of Pythagorean.

Music is describable in solely mathematical terms, but it also involves materials, the transmission of sound through air, and the psychological activity of sense-perception. Archytas was not only interested in the mathematics of music but in what we would call the physics of music and our experience of it: the instantiation of mathematics in the perceptible, three-dimensional world. Thus he observed and perhaps even experimented with the different pitches produced by different lengths of reed. He put forward the argument that sound is due to impacts produced by the collision of moving bodies. High-pitched sounds, he suggested, travel more quickly to our senses; low-pitched ones are slower.[39]

In his comments on how to make catapults the mechanist Hero, possibly working from a text by the earlier mechanist Ctesibius, describes effective ones as having 'harmonious proportions': a term that harks back to this early Pythagorean discovery of music ratios and its corollary, that some ratios work better than others. The most famous Pythagorean conclusion to be based upon this insight was the idea that each of the circular motions of the moon, sun, planets and stars around the centre of the cosmos produced a distinctive note dependent on their distances from the earth and each other and their speed of motion. For instance, Pliny says that there was a semitone difference between the note of the moon and that of the planet Mercury (although other evidence suggests the mathematics involved was aspirational rather than having been actually worked out). In combination these notes harmonised to produce the so-called 'music of the spheres': an eternal natural orchestra.

And that was just one way in which some Pythagoreans extrapolated from their discovery of the mathematical basis of music to the theory that there was a similarly numerical basis for the rest of the physical world. Integers were thought to be fundamental to concepts and classes such as woman, marriage or justice (respectively the numbers two, five and four).

It is not clear whether Archytas himself subscribed to theories of a numerical physics or celestial harmonies. He does, however, seem to have articulated an argument that social justice within the city-state or polity was a matter of understanding ratio and proportion; and also, according to one later source, suggested that inequality is the cause of motion (Simplicius, *On Aristotle's Physics* 431.4). A world out of balance is a world in continuous flux: pulled between countervailing tensions.

The notion that there is a real fairness or proportionality in the scheme of things and that mathematics is a means of discovering it remains, but modern research takes into account the fact that value is subjective and may vary. The solution sought is a 'proportional envy-free allocation protocol', and cake often serves as the metaphor of case study for a divisible item of which people might prefer different parts.

An envy-free protocol was worked out in the 1960s for a situation with three recipients. Today, in the EP (Equitability Procedure) for sharing out something among three people or more, the first step is for the recipients to tell the cake-cutter what part of what's on offer that they value most (the icing or the filling?). Each recipient then gets an equal amount of the part they value the most and the remainder is shared out proportionate to the value they each gave it.

So far so good, but the solution becomes increasingly complex as more people get involved. The Brams-Taylor protocol is already lengthy for just four people, involving 20 steps of which one is itself a long sequence of decisions and choices by the recipients in turn. To extend this into a solution that can be used for any number (n) of recipients, you employ the formula $2^{(n-2)}+1$, which gives you the number of pieces to be divided among those n recipients. So if there are to be seven recipients, there are already 33 pieces; if 12 recipients, 1,025 pieces. This also requires a situation in which all the recipients are able to be interrogated about their preferences and it is tamper-proof only in the sense that if people get their preferences wrong they may be dissatisfied with the end result. That might work as a deterrent against lying, but does not allow for being mistaken or changing one's mind about what one's preferences are. Imagine states instead of individuals, trying to divide land or water resources between each other (many rivers flow through multiple countries). We can guess that a mathematical protocol, although fairer than most other solutions, would not be immune to human error and discord. The preferences of all recipients can be given equal mathematical weight, but some would have more economic or military power than others.[40]

In Greek thinking, the idea was less that there is a mathematically equitable way to distribute items of subjective value and more that balance and harmony were the cause and structure of things that people found good, from a well-tuned lyre to a fair distribution of goods and duties. This view of the universe became deeply ingrained in much of Greek scientific and philosophical thought and its power as an explanatory trope found its way into aesthetic, civic and political discourses as well.

Although the Pythagoreans took the explanatory possiblities of actual numbers to an extreme, the fact that mathematics mapped onto the real world and could be used to predict and exploit it suggested to many technical specialists and to the intellectually adventurous in general that nature – including artefacts that exploited natural behaviours – was fundamentally mathematical; that physics, biology and society expressed mathematics and, therefore, that mathematics could be applied to sculpture, architecture, artillery and politics.

Physicians like Erasistratus began to interpret the body through mechanistic principles (Chapter IV) and Erasistratus' contemporary Herophilus measured pulse rhythms with a water clock, itself a form of technology that had been much improved by Ctesibius and other mechanists using their experiential understanding of fluid dynamics. It enabled Herophilus to put into practice a conviction that the human body could be quantified and described in terms of numbers and rhythm, from the length of its duodenum (from Herophilus' coinage *dôdedaktylon*,

12 fingers long) to the specific musical rhythms he thought normal in the pulse at each human stage of life (child, youth, prime, age).

One of the most famous categories of mechanical invention were the automata, model or life-size imitations of animals, people and situations that moved and made appropriate noises without any apparent external cause even at the beginning of their motion. The philosopher Sylvia Berryman has suggested that these self-moving images of nature reinforced or even began the concept that natural things could themselves be mechanical: a system of cause-and-effect that gave only the appearance of direction or intelligence.[41]

To do mathematics was therefore a way of doing natural philosophy, particularly as Greek mathematics developed an approach that placed deductive proof and a consequently high level of specialist consensus at its heart: a way of doing thinking for which mathematics was the model and ideal. From this perspective, then, mathematics and mechanics could rival philosophy's claim to be the best path to truth and happiness – the discovery of the right way to live, which Archytas had argued was a matter of finding out the mathematics of how society worked. One example of such a rivalry can be seen in Hero's discussion of – again – catapults and other war machines, and their implications for social and personal philosophy.

Weaponise Your Way to Inner Peace

The largest and most essential part of philosophical study deals with tranquility, about which a great many researches have been made and still are being made by those who concern themselves with learning; and I think the search for tranquility will never reach a definite conclusion through the argumentative method. But mechanics, by means of one of its smallest branches – I mean, of course, the one dealing with what is called artillery-construction – has surpassed argumentative training on this score and taught mankind how to live a tranquil life. With its aid men will never be disturbed in time of peace by the onslaughts of enemies at home or abroad. Nor will they ever be disturbed when war is upon them, because of the scientific skill which artillery-construction provides through its engines [...] After peace has continued for a long time, one would expect more [peace] to follow – when men concern themselves with the artillery section, they will remain tranquil in their consciousness of security, while potential aggressors, observing their study of the subject, will not attack. But every act of aggression, even the most feeble, will overwhelm the neglectful, since in their cities artillery preparation will be non-existent. (Hero, *Artillery Construction*, 71-3[42])

This is the opening to a treatise on how to construct ballistic war machines by the mechanist Hero of Alexandria (AD 10–70), drawing on work over the previous two or three centuries. The Greek word translated here as 'tranquility' or 'tranquil life' is *ataraxia*, which more literally means something like 'freedom from disturbance'.

As Hero begins by pointing out, *ataraxia* was an important philosophical goal among intellectuals of the period. The philosophical schools of thought that emerged over the third to first centuries BC, chiefly the Stoics, Epicureans and New Academics (also called Sceptics), argued that their way of thinking, carried out properly, produced such tranquility in the individual.

The appeal of *ataraxia* was that it represented freedom from fear and immunity to disaster or difficulty. The argument was that avoiding life's perils is almost impossible and certainly unlikely. Illness, war, natural disasters, bad weather, shortage of food, loss of work, political dangers, crime, death or illness of family and friends, rising expenses, housing shortages, marital or other family troubles. The list is familiar to us, but in the ancient world such dangers were usually both more likely and more grievous.

Philosophy suggested that if you could not avoid such omnipresent hazards, you could learn, in essence, not to mind. Instead of controlling the external world, you could control your response to it. This would also mean that you ceased to fear disaster that had not yet occurred and only might.

The different philosophical schools of thought offered different ways in which to do this. For the Stoics, the world is determined by god for the sake of what is good, so that everything that happens to you is both inevitable and for the best, in the best of all possible worlds.[43] Once you understand that, you will not only be resigned to fate, but happily resigned, since all you are actually suffering from is a kind of epistemic shortfall. A collapsing pension fund, painful arthritis or imminent invasion are only dreaded because you cannot see their role in the larger picture. The Stoic view is that the only thing in life which is really good or bad is one's own character: unlike virtue, wealth and health are really a matter of indifference. You can prefer them to poverty and illness, but only in the way that one might prefer strawberry to chocolate ice cream.

This approach has obvious similarities with many religions, in that it denies meaning to most worldly things and asserts in their place a trust (the Stoics would have said a proven rational understanding) in a larger purpose. All the individual has to do is realise this and resign themselves to it.

The Epicurean view was somewhat different. It placed considerable emphasis on rejecting the existence of traditional gods and the afterlife as incoherent and unnecessary beliefs. Eliminating these, in favour of atomic explanations for life, the universe and everything, also eliminated a great deal of the fear prevalent in peoples' lives and helped them realise *ataraxia*. As for remaining problems, the Epicurean explication of hedonism concentrated on maximising pleasure by minimising pain, thereby achieving an overall positive effect. Friendship, for instance, was a pleasure sufficient to counteract disease and political strife and stress could be avoided by withdrawing from public life. There was no need to fear alcoholism, accidents or a large wine bill, since the Epicureans advocated moderation. As with the Stoics, the essential theme here is that you can adjust yourself to the universe by reforming your perceptions, beliefs and life.

Even the Sceptics suggested that their own position, one of carefully maintained epistemic non-commitment, could lead to *ataraxia*. Since much of life's anxiety comes from uncertainty over what is happening or about to happen, and what this might mean for other things, the Sceptic can argue that once one realises that certainty appears to be impossible, matters become much simpler and possibly imaginary. At the bare minimum it allowed one to avoid any emotional commitment in struggles over politics, religion or philosophical theories. Life washes over the Sceptic, who is not quite certain that either he or it is there.[44]

At this point, consider Hero's remarks again. Hero raises the philosophers' goal of *ataraxia* by dismissing their claims to be able to solve it: they have not, he says, and they probably cannot. He would have had a point: success at achieving *ataraxia* by any of these very different methods looks both difficult and far from certain. Fortunately, Hero has an alternative to propose. Build good quality artillery!

The incongruity of comparing artillery construction to the elaborate, fiercely argued, all-encompassing theories of the philosophical schools is probably deliberate. It may well be a joke, but it is a joke at the philosophers' expense, and it contains a serious attempt at an alternative worldview, not just in regard to *ataraxia*, but about how we should go about understanding the world around us. The claim that tranquility comes from living in a city with good artillery is a claim that peace of mind comes from controlling dangerous external factors rather than your reaction to them – a reversal of the philosophers' approach. Hero does not offer this suggestion just as an alternative to philosophy, on the grounds that inner peace is not for everyone. He dismisses philosophy as a solution to this problem at all ('the search for tranquility will never reach a definite conclusion through the argumentative method') and asserts that mechanics has 'surpassed argumentative training on this score'.

This assertiveness is a challenge to the preferences of Greco-Roman elite culture, in which philosophy was considerably more prestigious than mechanics. Philosophy valued conclusions and proofs that could be expressed and preferably proven in logical or mathematical form, rather than in arguments that depended upon empirical evidence or involved actual material things. Although some strands in philosophy emphasised empirical research as a means of getting to the truth, even Aristotle thought that conclusions arrived at through induction should be confirmed by deductive logic. This was one reason why mathematics was to many philosophers an ideal form of knowledge, because mathematical theorems were abstracted from particular items and were consistently true, making them eternal. For some thinkers – influential individuals like the Pre-Socratic Parmenides and subsequently Plato – the material, experiential world was a deceptive set of illusions, which had to be discarded before reason could derive eternal truths from first principles. Mechanics in contrast emerged from later Aristotelianism. Its theoretical underpinnings in natural philosophy came chiefly from Strato of Lampsacus, known as 'the physics man', who was the third head of Aristotle's Lyceum and went on to become a courtier and royal tutor at Alexandria.

One aspect of this cultural preference for abstract thought over experiential information is socio-economic. Making things, finding things out by making things or even thinking of made things as epistemologically and ethically valuable, was associated with manual craftsmanship and labour. These were socially lower-status activities compared with the land-ownership and farming of aristocrats.[45] Philosophy was usually an activity of those who did not have to work for a living, let alone with their hands. Working for payment or survival entailed a lack of autonomy that Greek culture interpreted as a form of slavery.

Aristotle divided kinds of knowledge into the theoretical, pursued for its own sake, such as metaphysics and the natural sciences; the practical, with the aim of social and individual good conduct, such as politics and ethics; and the productive, the creation of beautiful or useful objects (a wide range of subjects from medicine to ship-building and music).[46] The division was hierarchical: theoretical knowledge was the highest form of activity. Mechanics and artillery construction, however dependent on maths, would have been part of the lowest division, productive knowledge.

In sum, mechanics was perceived by the intellectual elite of the Greco-Roman world, who also tended to be its wealthier and more leisured classes, as an inferior form of knowledge carried out by an inferior class of people for inferior reasons. Biographical anecdotes (of dubious veracity) about the important early mechanist Ctesibius of Alexandria identify him as the son of a barber. This is not intended as a compliment.

This is the cultural context in which Hero is arguing that mechanics, not just theoretical mechanics but the productive science of artillery construction that was aimed at building better and more effective machines for paying patrons, is a more successful form of knowledge than philosophy. The fact that the mechanists' creations actually work, while philosophers merely argue, suggests that they have a better claim to knowledge. In addition, he claims, this kind of knowledge can actually succeed in producing happiness or tranquility.

Hero is implying that practical knowledge is just as good, if not better, as th e purely theoretical kind. Philosophers may have found this claim outrageous. Modern philosophers also tend to react badly to any suggestion that science has superseded their method of inquiry, such as by explaining religion as a neuroscientific or evolutionary phenomenon, or doing ethics with experimental psychology, or asserting that consciousness can be explained reductionistically. They sometimes argue that philosophy deals with the knotty parts of problems where science cannot go and emphasise that questions are as important as answers.[47] Scientists, who are largely engaged in looking for answers, are, like Hero, not always impressed by such claims.[48]

There is one further point to make about this passage of text. Hero is thinking in terms of the actions and beliefs of city-states – the standard political community of the Greco-Roman world. This plan for tranquility through force of arms is a civic one for the community as a whole, rather than an individual's personal achievement. Similarly, the potential attacker is discouraged as a society, not as an individual.

Artillery construction is necessarily embedded within such civic and political structures, while philosophy is more independent. Numerous anecdotes about philosophers or doctors or artists narrate them defying kings or other sole rulers: the mobile intelligentsia of the Greek world had an ambiguous relationship with patronage. Hero therefore configures that need not as a patron – client relationship, but in terms of civic identity and strength – something traditionally valued by Greek and Roman culture.

Hero's particular argument about the civic virtues of artillery construction is an early theory of military deterrence, made on the basis that an easy target is an invitation to attack, while a well-defended city or state will be left in peace. This is the familiar assumption of any defence (or war) department and appeals strongly to that dubious arbiter, common sense. By going so far as to suggest that good technology guaranteed peace and consequently peace of mind, however, Hero left himself open to counter-examples from history. These are not difficult to find. A famous example would be the siege of the coastal stronghold of Tyre by Alexander the Great and his Macedonian forces in 332 BC. According to the detailed account by the later historical compiler Diodorus of Sicily, the Tyrians had been early adopters of the Heronian School of Defensive Thought:

> They had plenty of catapults and other machines useful for sieges and they easily constructed even more because of the machine-builders [mechanists] and other technical experts of all kinds who were present in Tyre. (*Library*, 17.41.3)

This did not succeed in deterring Alexander (very little did). He took Tyre after a seven-month siege. Diodorus attributes this largely to Alexander's 'love of glory' but ensuring that a powerful naval port was not left unconquered to his rear probably had something to do with his persistence. The episode does demonstrate that effective defences make attack difficult and would therefore put off or defeat lesser attackers, but a philosopher could say that the mechanists' path to tranquility is no more of a guarantee than theirs. There is always a risk that the enemy has a bigger catapult or a more ingenious technical expert or some other way to achieve victory – starving the besieged, for instance.

What Hero could not articulate, because the technology did not exist, is the nuclear variation on the theme of deterrence. The Cold War doctrine of Mutually Assured Destruction (MAD) marked a step-change here, since once two entities had the capacity to launch nuclear weapons at each other, which of them had more or better ones became largely irrelevant – not that this logic has prevented wealthier states from accumulating more and improved missiles. Only an effective mechanism of interception (as yet unachieved) or sabotage (computer viruses may succeed here) would shift nuclear warfare from a zero-sum to a winnable, if risky, game.

As far as I know no-one has ever claimed that MAD produces a feeling of tranquility, but perhaps they should. It is arguably more likely than Hero's 'build a better catapult' to actually maintain a state of uneasy peace. Current events, on the other hand, suggest that Hero's belief in cities with no weapons being more likely to be attacked is also a belief of many states without nuclear weapons, who react by trying to acquire them. Yet for some reason worldwide MAD does not seem to be anyone's favoured option. Perhaps, on balance, one should stick with Epicureanism in the search for tranquility.

CHAPTER VI

THEN AND NOW

On Not Being Human

The physician Galen, one of the ancient world's leading proponents of anatomical investigation and its usefulness in medicine, regretted the fact that he could not do much in the way of human dissection, being restricted instead to animals: chiefly dogs, pigs and the Barbary ape (macaque). He recommended that students of anatomy make use of bodies washed out of graveyards by floods or the victims of brigands, left by the wayside.

Human dissection was not often, so far as we know, explicitly forbidden by ancient governments, but a general reluctance to indulge the few physicians interested in obtaining corpses meant that they were difficult or illegal to obtain. One exception to this was the situation in Alexandria, the newly built capital of Greek-ruled Egypt, in the mid-third century BC. There the Ptolemaic dynasty of Greco-Macedonian kings offered patronage to all kinds of investigators into the natural world. Among these the physicians Herophilus and Erasistratus – the former born in the Greek colony of Chalcedon near the Black Sea and the latter from the Aegean city and island of Ceos – dissected both animals and humans. They worked separately and their own works are almost entirely lost but later authors, like Soranus and Galen, discussed, often in detail, their findings and theories. Much of their work remained controversial, but some aspects – like the anatomically based proof of the brain as the organ of thought, the structural differences between veins and arteries and discoveries of such important anatomical features as the heart valves or Fallopian tubes, not to mention the anatomy of the eye – became well established among physicians interested in such things. Only Galen had a similar or greater impact on ancient anatomy.

Herophilus and Erasistratus are also the two names associated with a practice even more controversial than the dissection of dead humans: the vivisection of live ones in pursuit of anatomical knowledge. Animal vivisection was not so rare. Galen and others utilised it, as have many other researchers in subsequent medical history. But there is good evidence that the two Alexandrian anatomists, although they used animal vivisection, also went further than this.

The principal evidence comes from a text written much later: the medical encyclopaedia by the Latin author Aulus Cornelius Celsus (*c.*25 BC–AD 50). It was part of a larger encyclopaedia of all topics but the rest of it no longer survives. We know very little about Celsus himself. He might have been a doctor, perhaps with the Roman army. Certainly he knew a lot about the history and contemporary nature of medicine and – like all ancient authors – was in the habit of expressing his own opinions on difficult or contentious issues.

In his introduction to Book 1 of his medical survey, Celsus describes the origins of Greco-Roman medicine and discusses past and contemporary methodology:

> Moreover, as pains, and also various kinds of diseases, arise in the more internal parts, they hold that no one can apply remedies for these who is ignorant about the parts themselves; hence it becomes necessary to lay open the bodies of the dead and to scrutinize their viscera and intestines. They hold that Herophilus and Erasistratus did this in the best way by far, when they laid open men whilst alive – criminals received out of prison from the kings – and while these were still breathing, observed parts which beforehand nature had concealed, [...] Nor is it, as most people say, cruel that in the execution of criminals, and but a few of them, we should seek remedies for innocent people of all future ages.
>
> (*On Medicine*, introduction to Book 1, 24–7)

Although this is a piece of evidence much later than the events it mentions, Celsus and his contemporaries clearly accepted it as fact – the dispute was over whether the practice was appropriate, not over whether it had happened.[1] These contemporaries included physicians who identified themselves as 'Herophileans' and 'Erasistrateans', that is as members of what were still, in Celsus' day, two of the leading schools of thought on physiology and clinical practice. Celsus' brief paraphrase includes circumstantial details which suggest that even fuller accounts were available. At the time writings by Herophilus and Erasistratus themselves were still in existence, as were treatises by their students and rivals from the succeeding generation. Moreover, some of the Alexandrians' discoveries in anatomy, notably the distinction between the motor and sensory nerves, would have been very difficult to obtain if they were not performing human vivisection.

What we do not know is the immediate context: whether or how Herophilus and Erasistratus justified this step from animal dissection and vivisection to human dissection and vivisection; or what their contemporaries thought of it. Alexandria in the third century BC was an unusual environment. It was only 50 to 60 years since it had been built as the Greek capital in Egypt and, unlike the Greek city-states or even the Roman republic, it was the capital of an absolute monarchy. Herophilus and Erasistratus belonged to a very small and itinerant intelligentsia as part of a socio-economic elite that at the time, in both Egypt and the wider Greco-Roman world around the Mediterranean, was composed largely of Greeks or people who could speak and think in Greek. Alexandrian intellectual society was geographically located within Egypt and had a complex two-way relationship with the older culture of its majority inhabitants, but its Ptolemaic kings were just as interested in Greek cultural influence. Patronage of Greek intellectuals was their weapon, and part of what they could offer was simply power – a power to do what in other social contexts was unthinkable.

Celsus says that the vivisection subjects were 'criminals received from prison by the kings'. Criminals have a low status in any society and in antiquity punishments were frequently harsh. The death penalty and torture were both possibilities, although there were restrictions on the class of person and crime to which these applied. Such punishments re-categorise people as subjects to whom normal rules and values no longer apply. In a sense, they dehumanise them. This attitude is clearly present in the pro-vivisection argument that Celsus reports: criminals are contrasted with the innocent people their coerced sacrifice will save. In this equation, they count for less.

The distinction was a sharp one in the Hellenistic kingdoms like Alexandria, where society was already very stratified along lines of power and status. The king became divine after death, as did, later, Roman emperors. Criminals – probably ill-educated, possibly Egyptian rather than Greek, expelled from normal society and stripped of legal status – verged on being a different species. Histories written during this period and within the next few centuries accuse other Greek kings of the time of testing poisons on another category of people of very low status: slaves.

Given all this, the most surprising thing about the vivisection issue, as reported by Celsus, is that part of the contemporary debate was about ethics as well as about whether vivisection was methodologically useful. (Celsus' own conclusion, by the way, was that dissection is an appropriate method but that vivisection is unnecessary.) The precise words, from the quotation above, are: 'Nor is it, as most people say, cruel that in the execution of criminals, and but a few of them, we should seek remedies for innocent people of all future ages.'

This is partly an argument about ends justifying means; in this case, that a greater total of lives saved (in the future) justifies a smaller number of lives lost now. As we just saw it does not count the lives in each column as the same kind of life. Criminals are contrasted with the innocent, as well as the few with the many. As its opponents probably pointed out, this argument has to assume quite a lot. In particular, it assumes a fact not yet in evidence – that its programme of investigation will be successful in discovering reliable knowledge and useful treatments. A subsidiary assumption is that a 'few' criminals will be enough to produce this information, enabling them to be weighed against the hypothetically saved millions of the future. It is not clear from Celsus' summary of the argument whether only the (by then long-dead) victims of the Alexandrian physicians are under consideration or whether another few test subjects could be included if human vivisection became a practicable and patronised methodology again.

(Celsus also mentions arguments against anatomical investigation from the other principal schools of medical thought in his lifetime, the Empiricists and Methodists; arguments well-known from other sources. They are directed against dissection itself, whether animal or human, living or dead. The main thrust is that the very act of opening the corpse is a change that might have had an effect on the internal arrangement of the body – the investigator has no baseline from which to compare. In any case, dissection was by definition already an investigation of a non-working system, and vivisection of a traumatised, injured one.)

The accusations of cruelty mentioned by Celsus as the opinion of 'most people' are clearly specific to (human) vivisection. Galen's dissections and vivisections of animals were performances that drew crowds, although he did advise against using macaques as subjects for vivisection of the brain because the audience apparently did not like the expression on the monkey's face during the procedure.[2]

In any case human vivisection remained an isolated event in Greco-Roman history, but it is less clear whether the fatal handicap was a general dislike of the procedure or the fact that its usefulness was highly contested. Human dissection, as we have seen, was also rare. Aristotle, centuries earlier, had had to defend the unpleasantness of even animal dissection through appeal to the beauties of nature's blueprints that it revealed.

Celsus' own careful positioning on the issue suggests that dissection (probably combined with animal vivisection) was more acceptable than human vivisection, even among those who believed – unlike the Empiricists and Methodists – in the scientific and medical usefulness of anatomical investigation. But the moral appropriateness of such procedures and categorisation was contested and socially undefined. Human vivisection occurred in an era without a concept of universal human rights: instead, hierarchical social categories and moral judgments made being (a worthy) human a less biologically definitive affair.

Defining worth is at the heart of the controversial status of human vivisection in antiquity. If we take away the argument about whether vivisection can achieve medically useful knowledge, what is left is an argument over whether 'a few criminals' are worth the loss of 'innocents of future ages'. Structurally, the argument is identical to the modern debates over animal experimentation and vivisection in medicine. There is even an epistemological component to this controversy as well. Important treatments have been discovered or tested in animal trials – insulin, for example; or the heart–lung bypass machine. There are also a large number of pharmaceuticals or findings that have succeeded in animal trials (often performed on mice), but have gone on to fail in, or be inapplicable to, humans; possibly because of physiological and genetic differences between species or because of publication bias among experimenters (weak results are not reported).[3] The heart of the matter, however, is the claim that one category of organism (criminals; animals) is not of equal value with another (innocents; humans) and hence that the first can be justifiably used to improve the lot of the second.

Human and animal is not the only division in play: in recent years people including scientists have argued for the exclusion from experimentation of the great apes. Different scientists draw different lines. To take one random example, the neurologist Steven Rose uses day-old chicks (in dissection not vivisection) but would not use dogs in the same way.[4]

Part of the problem is that modern definitions of what I am calling 'worth' (since it is not an objectively measurable criterion, but depends on what individuals and cultures find of value) are not identical: some regard intelligence as important, some self-awareness or the related concept of consciousness. Consciousness is not easy to define and intelligence (also hard to define) is something that one has more or less of, rather than either having it or not. How much is enough, to count as a person?

Given human history it is always unwise to take anything for granted, and there have been incidents as recently as the twentieth century of governments and military institutions using particular groups of people in medical experiments without their permission. The Nazis are the obvious and worst example, but not the only one. But the same difficulties that beset the ethics of this question 2,000 years ago complicate our decisions now. It is not just animal experimentation that raises uncomfortable questions; justifications for torture also use the argument that bad people and innocent ones are not equal. One of these organisms is not like the other.

Science has not, and cannot, supply these answers because society is not sure what the question is. A biological definition of being human leaves out moral dimensions or social value, and ignores what humans have in common with other animals. We make judgments about the relative worth of individuals and groups all the time, but we often do not think about the presuppositions on which our judgments are based.[5] If a discussion about the merits of human vivisection seems part of a thankfully distant past, lurking in an ancient and obscure work of medical history, the familiarity of the arguments can provide an unexpected perspective on our own concerns.

Tomorrow's Pseudo-Science

One feature of life in antiquity has survived to our own time more or less intact. Then, it was a scientific theory; now it counts as 'pseudo-science'. It is of course astrology, in particular the science of horoscopes.

A horoscope predicts significant things about a person's life on the basis of the date and time of their birth, because mathematical astronomy could reliably work out where the sun and planets were in relation to the constellations of the zodiac at any given time.

From the earliest times, the Greeks knew that five stars did not behave like the others: Mercury (Hermes), Venus (Aphrodite), Mars (Ares), Jupiter (Zeus) and Saturn (Chronos). They called these the 'planets', which means 'wanderers', because they moved in a semi-erratic path around the celestial sphere against the slower and regular rotation of 'fixed' stars in their patterns of constellations. The planets could be seen as moving – changing position – against this background: they shifted east, paused, moved backwards west and then moved eastwards again. A great deal of the impetus of Greek astronomy came from efforts to explain this anomalous behaviour in a mathematically elegant way.

In spite of their stops and starts, however, there were regularities in the planets' movements. In particular they never went far from the path of the sun across the heavens: the ecliptic. As later ancient astronomy knew, the narrow circular band in which the planets moved was between 8 and 9 degrees north or south of the sun's ecliptic. This meant that the wandering stars were only ever seen against and moving through a particular group of constellations. In addition, the path of the moon was in the same zone.

The Greeks were far from the first to notice this. Around the seventh century BC Babylonian observers of the stars divided this ecliptic band into 12 longitudinally equal sectors, each with one constellation. The time taken for the sun to pass through this sector of sky – through each constellation – was a month.

The constellations in question were those of what Greeks called the zodiac, from the Greek words for life and living animals (*zoê, zôios*), because almost all of them formed the images of animals, humans or mythological creatures. By about the time that Meton was devising his 19-year solar calendar in 432 BC, the Greeks had access to Babylonian star-catalogues and records, including this early form of cosmic mapping. Each of those 12 sectors of the ecliptic path formed 30° of a 360° circuit of the sky.

The Babylonian motive for developing this reference system for the positions and movements of the stars, and their identification of the fixed stars of the ecliptic zone as particularly important, was divination. Predicting the future through signs can take many forms: the Greeks and Romans, for example, both used the entrails of sacrificial victims to predict the future (haruspicy). A specialist college of priests, the augurs, analysed signs of the gods' will on behalf of the people and Senate of Rome, using a variety of information including the local environment and the behaviour of people and animals. Sacred chickens were taken on Roman campaigns; their actions, as interpreted by the augurs, signaled the celestial odds on the next day's battles. Unusual events also usually counted as omens, whether these were agricultural (two-headed animal births), meteorological (rains of blood) or astronomical. In Thucydides' *History of the Peloponnesian War*, the Athenian general Nicias interprets a lunar eclipse as a warning not to retreat from Sicily. Staying put, he and his forces became besieged, with disastrous results.

There was scepticism about divination. Hippocratic and later texts declare that while some dreams are plainly sent by the gods, others are purely internal affairs. Thucydides' account of the Nicias affair does not endorse Nicias' interpretation of the eclipse or the notion of omens as valid. In Plutarch's (much later) biography of the Athenian politician Pericles, the seer Lampon and the Pre-Socratic thinker Anaxagoras offered competing interpretations of a ram that had only one horn, growing out of the middle of its forehead. The seer divined a political forecast: out of the two rivals for political power at the time there would only be one winner – and that would be Pericles, whose ram it was. Anaxagoras dissected the ram's skull and showed how its internal structure matched the ram's horn, suggesting that the horn was just the result of natural chance. According to Plutarch, 'it was Anaxagoras who won the plaudits of the bystanders; but a little while after it was Lampon, for Thucydides was overthrown, and Pericles was entrusted with the entire control of all the interests of the people.' (*Life of Pericles* 6.2–3. The Thucydides mentioned here is not the historian but a different politician.)

Thus there was throughout the Greco-Roman world and societies around the Mediterranean a widespread if not universal belief in the possibility of forecasting the future and a number of individuals and techniques who were formally or popularly associated with this task. Astrological forecasting emerged as a hybrid technique of Babylonian stellar divination and Greek astronomy in the second and first centuries BC, apparently in Greek-ruled Egypt.

The earliest horoscopes extant in Greek come from the mid-third century BC, but are much more common from the late first century BC onwards. Predicting the course of someone's life based on the planets' relations to the stars at the time of their birth rapidly became a popular mode of divination in Greco-Roman society. Some authors, like the Latin satirist Juvenal, described it as appealing chiefly to women and the lower classes, but Augustus' successor as emperor, Tiberius, consulted an astrologer on political decisions and it was a well-established part of social intercourse by the first century AD. Its Babylonian origins were reflected in the use of the term 'Chaldeans', the name of the original caste of Babylonian omen-priests, to refer to all astrologers, but its success had a lot to do with the increased authority of astronomy, which Julius Caesar had used for calendar reform. The same observations and calculations formed the foundations of both disciplines – indeed, once astrological horoscopes had emerged, the two were often indistinguishable. Antiquity's most famous astronomer, Claudius Ptolemy, also wrote a major work on astrology, the *Tetrabiblos*.

Not all theories of how astrology worked (if it did work) were the same. For some people astrology was a more sophisticated rendition of the idea that stellar phenomena acted as signs from the gods. Ptolemy, however, had a more technical explanation for astrology, which relied on the mechanical and mathematical model of the universe developed by astronomers to account for the movements of the stars.

Ptolemy's argument was that the movements of the planets in relation to the fixed stars, together with those of the sun and moon, affect the sublunary environment, that is, they affect the weather and climate. That would account for the fact that the seasons and temperature vary with the time of year marked by the passage of the sun and moon across the sky and through the zodiac. The moon correlates with the tides. The universe of antiquity was a much smaller place than ours – much smaller than our solar system – and it must have seemed an obvious explanation of weather patterns. The fusion of ancient astronomy and meteorology is evident in the almanac known as a *parapegma* (Chapter V), in which weather forecasts are correlated with astronomical events.

If celestial bodies affect the weather, they must also affect those parts of a human environment that the weather influences, from the air we breathe to the soil that is watered by the rains and dried by the winds. As discussed in Chapter IV, the central idea of ancient medicine was that people's constitution is partially fixed by heredity – a kind of long-term adaptation to an area – but also by their daily environment: what winds touch them, what they breathe, what they drink, what they eat. If that is affected by the local weather then the planets' movements are part of what makes every human who he or she is. And not just as a matter of a person's health but of their material constitution generally – that meant their temperament.

Ptolemy's theory is essentially that what happens in one's life has a lot to do with personality. Someone we would call an extrovert and a risk-taker, for example, is more likely than another man to make a fortune, but he is also more likely to die at sea. (This example is Ptolemy's, and it reflects a society centred around the Mediterranean sea basin, where much trade was maritime and ships were highly vulnerable to storms.) Planetary influence caused environments that partially caused a person's character, which was one factor in what happened in the rest of their life.[6]

This kind of astrology is about probabilities, not certainties, and quite small probabilities at that. Its usefulness as a guide to life was also undermined by Ptolemy conceding that many astrologers ('Chaldeans') were frauds; that the older calculations utilised by astrological prediction were somewhat imprecise and that astrology was in consequence a difficult and uncertain skill, rather like medicine. He could have added that data on people's birth date, and particularly on their time of birth, were hard to come by and might sometimes be inaccurate.

Ptolemy's version of astrology is thus a weak one – much weaker than the confident predictions offered by most practitioners. There were so many confounding factors that the effects of planetary influence did not appear much stronger than chance and were nearly as difficult to predict. Ptolemy's astrology is a very 'soft science' application of the 'hard science' of astronomy, and his assumption is that improving the observational and computational power of the latter would go some way towards making the former more accurate. As we shall see in a moment, modern astrology has made precisely the opposite move in defending its credibility.

Ptolemy's very weak version of astrological causation did provide an effective defence to some common objections to astrology, even if it also reduced its attractiveness. (It's no use being told I have a 5 per cent extra risk of dying at sea over my lifetime when I'm trying to decide if I want a ticket for *The Titanic*.) Sceptics pointed out that many astrological predictions did not, in fact, come true, but such objections could often be answered contingently – you had an incompetent astrologer. Other objections found problems with the theory or with whole categories of prediction: why did twins not always die on the same day? Why was the position of the stars at conception and during pregnancy not as relevant as those at birth? The answers to these were not sufficient to convince those already sceptical. (There is a difference of minutes in the birth time of twins; the universe naturally ensured that the same influences operated at birth and at conception.) A professional sceptic, the philosopher Favorinus, pointed out that there might be more planets than had yet been seen, which would mean that astrologers were basing their predictions on inadequate information.

He was, it turned out, quite right; but Uranus and Neptune were not observed – or rather, not observed and recognised as planets – until the eighteenth and nineteenth centuries. Small perturbations in the orbits of the inner planets gave their existence away, but only to an astronomical world that post-dated the acceptance of heliocentrism and Johannes Kepler's laws of planetary motion. The discovery that astrology had been leaving quite a lot of the solar system out did not inflict a deadly blow on its credibility. Indeed, astrology has been resistant to the universe changing around it to the extent that, so far from being interchangeable with astronomical theory, it is now a kind of historical time capsule of the universe as we saw it in the first century AD. In addition to a larger universe, with more planets, going round the sun instead of the earth, with an entirely different operating system of physics, chemistry and climate, there is a difference in the positions of the earth relative to the constellations of the zodiac. The reason for this is the wobble of the earth around its axis of rotation (because it is not a perfect sphere), so that the line of that axis draws the wider end of a cone on the imagined backdrop of the fixed stars. The celestial north pole, drawn in another imaginary line from the terrestrial north pole to its intersection with the celestial backdrop, also shifts. For an observer on earth, this means that our position, relative to the path of the sun or to the celestial sphere, is not quite what it was.

The total periodicity of the cycle (a complete circuit of the rotational axis, often known as 'precession of the equinoxes' after one of its effects) is a little under 26,000 years, but the shift in our position can be observed over a much shorter timescale than that. That is just what Hipparchus of Nicaea did in the second century BC, when he noticed that the longitudinal co-ordinates of a star had shifted between his own time and observations made a couple of centuries earlier. Over greater amounts of time this becomes much clearer and it is now some 1,000 years since astrology was invented – on the basis of even older Babylonian observations.

In that system, there was pattern of stars contained in each 30° sector of sky around the ecliptic, covering the narrow zone north and south of the sun's course in which the planets and the moon also travel. But the shift in the earth's rotational axis relative to the plane of the ecliptic means that the original zodiac constellations are no longer in the same alignment with us. Astrologers have had to decouple the 30° sectors of sky from the zodiac constellations, since the sun is not 'in' the same constellation on 6 August 2015 as it was on 6 August AD 1. Then, it rose in Leo; in 2015, it rises in Cancer. Someone born on 6 August, then, has an official star sign of Leo but was actually born when the sun was in the Crab.[7]

Astrologers insist that this does not matter: the predictive and/or causal power of the signs is nothing to do with the constellation, only with the sector of sky that the sun is in at the time of someone's birth. As with Favorinus' hypothetical planets, apparently game-changing changes in astronomy have no impact on astrology in spite of the fact that the astronomical discoveries are uncontested by astrologers. It has become a closed belief system which immunises itself against disproof by severing connections with science – more precisely, with any theoretical or actual factor that might be susceptible to disproof. Like homeopathy, the main astrological argument is existential: it is here and it works and it is science's problem if it cannot figure out how (assuming a purely subjective reading of 'works').

The way we relate to astrology now is very different to its place in antiquity. Then it was an application of a scientific theory: a new contender in a marketplace that also took religious signs and omens for granted. Now the position is reversed: all our methods of predicting the future involve science or at least social science, from astronomy and meteorology to economics and psephology. Meanwhile scientific astronomy has changed out of recognition, leaving astrology behind as a persistent social construct. It is ironic that Claudius Ptolemy, a leading and influential mathematician and theorist of the ancient world, would be hopelessly adrift in cosmology if transported into the present, but could still make a perfectly good living as an astrologer.

Hippocrates in the Twenty-First Century

Sometime around the year 400 BC, a man writing in Greek composed a polemical treatise on the subject of what he called the 'disease mistakenly said to be sacred'. Another text from roughly the same time, late in the fifth century BC, argued that the human body consisted not of one substance but four humours (Chapter IV). Illnesses were caused by an imbalance in these humours. A third work, also from *c.*400 BC, described the bones of the human body and explained how to deal with fractures and dislocations, often criticising the methods of rival practitioners.

By about a century later, scholars reading medical treatises at the library of Alexandria in Egypt ascribed all these works to Hippocrates of Cos, a famous physician of the fifth century BC (*c.*460–377 BC) who was mentioned as such by the philosophers Plato and Aristotle.[8] Many other such works were likewise grouped together as being by Hippocrates. Today there are some 60-plus works in this 'Hippocratic corpus', a group canonised and partly constructed by the nineteenth-century edition of Émile Littré.

It was recognised early in the history of the corpus' reception that all the treatises it contained could not actually be by one person, i.e. Hippocrates.[9] The corpus includes divergent, sometimes conflicting theories and approaches; its works are now dated to a period of over a century and therefore are the work of more than one person. Some seem to be written by professional practitioners; others look like the work of interested intellectuals. Distinguishing the 'genuine' works from the falsely attributed, however, requires judgment as to what a genuine work should look like and unfortunately there is no text, nor any report from another source, that can be independently identified as being by Hippocrates and therefore serve as a basis for comparison.[10] The usual method of identifying genuine Hippocratic works is to look for texts which seem to articulate attitudes or ideas common to significant parts of the corpus; or texts which seem persuasive or impressive; or those which are not obviously anachronistic[11] and appear likely to date from Hippocrates' lifetime in the later fifth and early fourth centuries BC; and thence to locate others which seem similar or compatible to these most probably 'Hippocratic' works. The unfortunate fact is that all these approaches assume that Hippocrates did write at least some of the treatises in question and that what he wrote met later standards for medicine. In other words, genuine works were identified by how Hippocratic they seemed to be, when Hippocratic is defined as what is characteristic of the works of Hippocrates. If this seems circular, that's because it is.

The majority opinion, since the late twentieth century, is that it is impossible to conclusively identify any of these texts as being the work of the historical physician Hippocrates of Cos. Instead, it is largely through the most cited and influential treatises and ideas – *Epidemics*, *Nature of Man*, the *Oath* and *On the Sacred Disease* – and their interpretation within the subsequent medical tradition that the idea of Hippocratic theories and medicine as a coherent theoretical set was developed. In a sense these texts created Hippocrates, the 'father of medicine', rather than the other way around.[12]

But if 'Hippocrates' as an individual is a historiographical invention, by about 250 BC, when the Alexandrian scholars collected early medical texts and assigned them to his authorship, he was already the authoritative name within the Greek tradition of naturalistic medicine. Leading physicians of the time, like Herophilus, sometimes argued explicitly against Hippocrates' views (i.e. against views found in the Hippocratic corpus), suggesting that the textual corpus of Hippocratic views had established a mainstream, historical authority. Living physicians were partly defined by their relationship with this corpus – with the figure of 'Hippocrates' – whether that was one of agreement, revision or divergence. Several centuries later, Galen laid out and fortified his own position in various methodological and theoretical wars by utilising 'Hippocrates', his reading of Hippocratic texts, as an ancient authority. Truth in medicine, he implied, could be found by going back to the original, rather than in the dead-ends of contemporary debate. Galen successfully presented himself as the heir of ancient medical authority by returning to its supposed source.

Curiously, this rhetorical move remains a notable feature of medical explanation and sometimes of actual medical positioning in current, twenty-first-century writing and debate. A rapid trawl through the archives of the last ten years of *The Economist* and *New Scientist*, and one year of the *New York Times*, produced over 30 articles mentioning Hippocrates. Nor is the phenomenon limited to newspapers and popular science. Hippocrates also features in scientific journals when a historical reference point is required.

The reason for his appearance is clear when the article in question concerns some feature of the history of medicine, but these are a minority of the total.[13] Somewhat similar are cases where Hippocrates is introduced as the only historical reference point, apparently to illustrate either the antiquity of the subject under discussion or the continuity of a certain facet of medicine, as in this example from *The Economist*: 'Since time immemorial – or at least as far back as Hippocrates – novice physicians have been taught to smell patients' breath for signs of illness.'[14]

Presumably the choice of Hippocrates is partly as a kind of origin point, since there are in the Western tradition of medicine no texts preceding the Hippocratic corpus, and partly a self-perpetuating familiarity with his name. Occasionally, when an author is particularly concerned to stress the antiquity of an idea, older medical cultures are also referenced: a discussion of dietary regimen as medicine in the *New Scientist* referenced both Hippocrates and Ayurvedic medicine (from India) as, respectively, 2,500 and 5,000 years old in support of its remark: 'The idea that your diet can improve your health is an ancient one.'[15]

The implication is that the length of time an idea has existed gives it some validity, or as the *New York Times* put it: '"Let food be thy medicine, and medicine thy food," Hippocrates said 2,500 years ago. We've always known that some foods are better for us than others.'[16] Perhaps this is because it suggests that a different society also found the idea in question plausible, an argument strengthened in the *New Scientist* example by two different cross-cultural examples being given. Different, independent sources, this trope suggests, have arrived at the same conclusion. Indeed, if even physicians without modern medical resources thought this, surely it should be obvious to us (an assumption that overlooks modern medicine's familiarity, as part of its own history, with Greek views).

The Hippocratic text in question says that diet is medicine for one obvious reason: they did not have much else to offer. Drugs consisted largely of purgatives, which were often dangerous in large quantities. Surgery was not sophisticated. It was also the case that all material treatments, from purgative drugs to cabbage (a favourite of Roman folk medicine), consisted of plants, parts of animals or minerals. Occasionally these were exotics from overseas; many more were obtained by professional root-cutters from the woods and mountains of Greece. A large proportion doubled as normal foods but could be differently processed when used as a medicine. This meant that there was a much less obvious disjunction than in modern times between medicines and foods or herbs taken for nutrition: both consisted largely of things pulled from the ground. The Hippocratic text is not, then, drawing our modern distinction between diet and drugs; it is in essence saying that everything is a drug and should be thought of as such in the interest of achieving health.

This trope is not untypical of how Hippocrates is used for validation in modern medical literature (and popular literature on medicine), where he often appears in contested arenas. A Hippocratic aphorism tends to be at once pithy and medically vague to the point of being, as it were, bromidic: saying that diet has medical implications is not controversial. What is at issue is what exactly that means in terms of medical practice and the distribution of resources: to prevention or to treatment, to foods or pharmaceuticals. Hippocrates lends authoritative emphasis to a particular axis in such spreads of opinion. This is usually for a kind of general approach (maintaining health through diet is a good idea) rather than for specifics – few modern researchers quote 'Hippocrates' on the virtues of taking exercise just after eating, for example; but there are some instances where he is brought in on a more particular division of opinion.

The *New Scientist,* investigating the argument that fevers evolved to kill off infection and hence should not be lowered by antipyretics except at potentially fatal temperatures, says:

> The idea that fever can be beneficial dates to the time of the Greek physician Hippocrates, 2400 years ago. Ironically, it was the emergence of modern medical science during the mid-nineteenth century that led to fever being seen as harmful.[17]

Again, the implication of this historical comparison seems to be to lend support to a modern view, that fever 'can be beneficial'. As with Galen's readings of Hippocrates, it turns out that the ancient source of modern medicine, whether this was a mere five and a half centuries ago (for Galen) or two and a half millennia (for us) can turn out to be right where his more recent successors have gone astray.

But this interpretation requires a decontextualisation of the Hippocratic texts. They have been divorced from their original meaning within their contemporary intellectual and social framework. Galen's theory and practice of medicine was far more similar to his predecessors than is ours, but his presentation of the corpus is already highly selective and heavily interpreted. Ours is much more so, which is perhaps why Hippocrates' support is usually supplied only in the form of those brief gnomic quotes or summary paraphrases of single passages or texts.

Let us take, first, fevers, and then move on to the areas where Hippocrates is particularly often cited: diet, prevention and the holistic approach. We may then consider Hippocrates in his role as an ethical guide and exemplar, most notably in the often cited *Oath* and the injunction to 'do no harm'.

In antiquity, a fever is less a symptom of some underlying state and more a diagnosable condition in its own right, although it also supervenes on other illnesses. It can result from an excess of yellow bile, since this humour is characterised by heat and dryness and linked to the element of fire. The idea that fevers could be beneficial, as described by the *New Scientist* above and in medical literature on this topic, is partly based on remarks in the treatise *Coan Prognostics*, where the advent of a fever is said to 'resolve' convulsions in tetanus and other acute conditions (although not if the fever predates the convulsion, according to prognostic no. 350). The theory is that the sweating involved in certain kinds of fevers took the toxic humours or other substances causing the illness with it, although this was not always enough and purgatives or bleeding might be prescribed in support (*Regimen in Acute Diseases*). Similarly, nosebleeds, vomit, pus and menstruation were also helpful events and hopeful signs, as each involved an evacuation of toxic or excessive substances. It is then indeed the case that much opinion in antiquity, including some passages in texts of the Hippocratic corpus, thought of fevers as 'beneficial' in certain cases.

This is, however, an extremely limited parallel to the more modern analysis of fever as an immunological response to infection by disease agents – viruses or bacteria. The idea of infectious disease is lacking. The crucial point is the evacuation by sweating, not the raising of the body's local and core temperature. But more important is the fact that even in ancient medicine, fevers were not normally beneficial.

Fevers were central to ancient medical thinking, not as a useful response to infection, but as a large proportion of the category of 'acute conditions' (short, sudden, dangerous) as opposed to chronic injury or illness. Patients and physicians managed their expectations of fevers through the doctrine of 'critical days'. A patient's fever was believed to reach a crisis point at regular intervals and fevers were classified by this periodicity. In a tertian fever a crisis occurred on every third day, counted inclusively; in a quartan fever, every fourth day; in a quotidian fever, every day – i.e. a continuous fever. Types of fevers could combine to create a variable pattern of crisis days. At a crisis the patient's condition was thought to be at a kind of tipping point, liable to get either better or worse. This was the moment when an intervention by a physician could be most effective, as fevers in a non-critical stage were harder to affect. It was therefore crucial to be able to diagnose a fever periodicity accurately, for the sake of both prognosis and treatment.

The concept may have originated in observations of patients with malaria. Fever rises in tandem with a stage in internal parasite reproduction, at consistent two-day intervals for three species of parasite (*P. falciparum*, *P. vivax*, and *P. ovale*) and three-day intervals for a third (*P. malariae*).[18] In actual cases fever can also appear everyday, as it does in non-malarial infections generally, producing the Greeks' continuous or quotidian fever. But the predictive success of this observation seems to have resulted in the concept being greatly extended until, through the notion of different kinds of fevers combining with each other, almost any data could be fitted into a periodicity classification. It became a usefully flexible theoretical tool and provided the fundamental structure for thinking about fevers in antiquity. The Hippocratic texts known as *The Epidemics* are famous for the careful observation of symptoms on a day-by-day basis in individual case histories on the island of Thasos in the mid-fourth century BC. The author's or authors' goal was probably to improve understanding of fever periodicities and their relationship to other elements of illness (symptoms).

Our modern use of this complex and central field of ancient medicine is selective. The semantic network of meaning for the notion that fevers were sometimes beneficial has been stripped away. At best, the comparison with Hippocratic thought on fevers is only an indication that fevers are not obviously bad: that an alternative model is possible. It lends nothing to the modern scientific argument for the usefulness of fevers, because the model of disease and health, not to mention chemistry and physics, is entirely different.

Sometimes Hippocrates is invoked not so much because of the actual arguments made in the Hippocratic texts or other ancient works on medicine, but as an example and exemplar of what modern science perceives as valuable in its history. As an exemplar the Corpus fits more easily into the subject of medical ethics, where issues and solutions have changed less than theory or practice over the last 1,500 years. The most famous example of this is the Hippocratic *Oath* text. Many US medical schools use a modernised version of this at graduation, although it is not a legal requirement, and some of its tenets are often appealed to in discussion of ethics and privacy:

> Regarding that which he learns because of his profession, a doctor should 'remain silent, holding such things to be unutterable'. So said Hippocrates, and although doctors no longer take his oath, their patients' medical data are still confidential, protected by case law, the occasional statute (records on sexually transmitted diseases, for example, are extra-secure) and the General Medical Council, which can strike off loose-lipped doctors. (*The Economist*, March 2009).

The status of the ancient *Oath* is less clear than that of Britain's General Medical Council. The *Oath* prohibits abortives. At least, it prohibits physicians prescribing an abortive draught, but since much other ancient literature – including other texts within the Corpus – describe drugs or other methods for abortion and contraception, it is clear that the *Oath* was not some kind of universal or even common guarantee for ancient physicians. Its words also prohibit the physician swearing it from 'cutting for the (bladder) stone', which would indicate a rejection of an often risky procedure, either on general grounds of safety or as an acknowledgment by those taking the oath that this is not their area of expertise. The prohibition of an abortive drink may be the result of a similarly cautious approach. A little over five centuries later Soranus observed in his *Gynaecology* that abortives can be dangerous, especially in later pregnancies.

It may also reflect, as does much else in the *Oath*, a concern for moral reputation. The involvement of a physician with female patients represented an unusual and sensitive relationship in antiquity that had the potential to bypass the husband, father or other dominant male in the household. At its most obvious, this included a risk of sexual transgression, a risk hinted at by the *Oath*'s prohibition of a physician having sex with a slave of the household he is visiting. Anecdotes of imperial physicians conniving with sexually promiscuous wives to poison emperors are a trope of Roman historical literature, neatly combining two common fears about physicians' access to the innermost workings of a household.

But outsiders also threatened the authority of a head of household less directly, by providing access to alternative advice, drugs and people. Control over female reproduction was imperiled. Soranus has to assert that reputable medicine does not include providing abortives for reasons of concealing adultery or preserving physical beauty.

The *Oath* seems thus to be partly a public declaration of reputability through the repudiation of controversial techniques and moral dangers. Similarly, confidentiality was crucial to patients placing their trust in the physician and allowing him access to personal and familial secrets. This is indeed also the case now, since the structure of medical interaction similarly involves a carefully demarcated suspension of normal social prohibitions in favour of a professionalised, context-specific intimacy.

For once the citation of Hippocrates and the *Oath* in particular represents a genuinely parallel case between the old world and the contemporary.

Invoking the former, while presumably unnecessary as a justification for modern confidentiality, does the rhetorical service of emphasising the longevity and importance of this aspect of medical practice. The tension between this keeping of secrets and the social pressure on doctors to uphold accepted values or legal authority is likewise a feature of both antiquity and the modern world, although the details of both have undergone some change. Now such questions involve age rather than gender – the supply of contraception to underage girls without their parents' knowledge; or an obligation to report certain crime-related types of injury.

Yet another oft-cited dictum is 'do no harm'. This seems to derive from *Epidemics* 1.11, 'do good or do no harm'. The *Oath* says that the physician will do no harm or injustice to his patients, though this does not seem to be a statement about medical methodology.

In antiquity there was a popular view that physicians often did quite a lot of harm, summed up by the elder Pliny's claim that numerous Roman funerary monuments bore the legend: 'it was the crowd of physicians that killed me'. Physicians were sometimes prosecuted by their patients' families if a death supervened upon treatment. Hence the advice of Hippocratic texts to other physicians not to treat unless medical treatment appeared likely to be advantageous, in other words, not unless the prognosis was good.

A 'do no harm' approach in antiquity is thus partly about saving the reputation of the practitioner. But a genuine parallel does remain, because 'do no harm' is a shorthand for reminding physicians of the risks of treatment, either in itself or in sacrificing quality for quantity of life. The *Epidemics*, if not the *Oath*, may well have expressed a similar scepticism about medicine's limitations. In this, as in several other examples of Hippocrates' aphoristic afterlife, a very general point about medicine is made through use of a convenient historical story or rather, through a long-established catchphrase.

Finally, the largest category of Hippocratic validations in modern medical literature is the discussion of diet, prevention and what is sometimes called the holistic approach to medicine.

At first sight, it looks to be a promising field of comparison. Much of naturalistic and indeed religious Greco-Roman medicine did indeed emphasise prevention through the regulation of diet and exercise. Medical texts asserted the importance of considering the patient as a whole and tailoring diagnosis and treatment to suit. Factors of age, gender, bodily type, local environment, diet, exercise and occupation were all relevant. As with issues of prevention versus cure and pharmaceuticals versus lifestyle, this seems to make Greco-Roman medicine a classic example, even an exemplar, of a kind of medicine increasingly advocated in modern society.

As usual, however, it is more complicated than that. Some of the factors producing the ancient emphasis on prevention, diet and a holistic approach to the patient do not apply to modern society and do not make good reasons for adopting such an approach. Like many choices about how we approach health and treatment, the Greco-Roman system is driven as much by circumstance, possibility and culture as by a logical, ethical response to objectively perceived truths.

In Hippocratic and later Greco-Roman medicine, health depended on the internal balance of natural substances, usually humours. As we saw in Chapter IV, an excess or lack of such naturally occurring internal substances and their associated qualities of hot, cold, wet and dry, produced a state of illness. Health was about staying within such normal parameters, although these varied between both populations and individuals.

Individual variations in constitution and predispositions to certain diseases, while due largely to inheritance and location, were also influenced by personal choices. Foods, partly as a result of the environment of soil, water and air they were grown in, and partly because of their intrinsic nature, had differing levels of hot, cold, wet and dry. These qualitatively affected the body which incorporated them. Waters were also believed to vary in such qualities, although water was in any case usually drunk mixed with wine. Air might be sodden with moisture or drying, hot or cold. One treatise advised breathing less.

So the condition of one's body and mind could be affected by diet. It could also be altered by activities such as exercising, which was thought to dry the body up and expunge moisture, as in sweating; or by bathing. As a group, these methods were called *diaitia,* meaning way of life or regimen – a more inclusive term than its modern descendant, diet.

Medical recommendations on such topics comprise a very large proportion of extant medical literature. Texts such as *On Regimen* and *On Regimen in Health* were often prescriptive to a precise and elaborate degree, involving not only what and how much to eat and drink, but when and in what order.

This conception of what constitutes health and disease identifies the patient themselves as largely responsible for their state of health. The role of inheritance is limited. External environment, especially the epidemic-inducing air, is the closest thing here to an external disease agent on the model of infection, but its effects can be rebalanced or guarded against by effective regimen. Health becomes a matter of continuous adjustment towards the optimal state of any and every individual.

The role of the medical practitioner, the physician, was largely to prescribe such a regimen, explained and justified through a theoretical framework of humours and qualities (or by alternative theoretical models, such as an excess of blood overwhelming the mechanics of the body in Erasistrateanism). Regimen, although it seems to have developed from the habits and requirements of daily life, became increasingly elaborate and precise as it evolved into the dominant form of professionalised medicine. Only those with independent means and consequent leisure, lacking other claims upon their time and attention, could have afforded many of its prescriptions. It was a conception of health and therefore of medicine that extended the role of the latter into all areas of life at all times. Illness represented a failure of adherence, of will-power and of morality: it was the reward of over-indulgence. In this, medicine both reinforced and exemplified a popular moral theme of Greco-Roman culture, the valourisation of self-restraint and the depiction of excess – ostentatious displays of wealth, promiscuity (male or female) and sexual appetite, gluttony, drunkenness – as a weakness which led inevitably to both bodily and mental degeneration.

The modern world includes similar tensions. Smoking and obesity are seen by many as the result of personal bad choices, while others emphasise addiction, genetics and environment as drivers of both lifestyle and its disease risks. Exercise is virtuous; thin is beautiful. Greco-Roman medicine is for us an ambiguous morality tale: it could just as easily be a story of blaming the victim or of medicalising eating habits, as it is one of prevention and healthy lifestyle.

Similarly, it seems probable that a large factor in this emphasis on regimen, on health, on prevention, is a consequence of the fact that there was not very much that doctors in antiquity could do about acute, dangerous conditions. Surgery for trauma, including some internal operations, developed considerably over the millennium or so of Greco-Roman history, reaching in some cases a high degree of technical and anatomical expertise. Galen, again, would be a leading example and more can be found in the medical encyclopaedia of Celsus or the later medical collection of Paul of Aegina (AD 625–90). But there was little that it knew to do against what we now think of infections, or against shock and blood loss. A large proportion of ancient surgery was carried out in non-acute circumstances. Operations included circumcision, cosmetic alterations and even the removal of cataracts.

Drugs consisted largely of emetics, laxatives, emmenagogues; as well as some anaesthetics and opiates. Many were poisonous in any quantity. They worked within the medical model being applied, in which toxic or excessive substances needed to be removed from the body. A similar rationale accounted for Greek physicians' use of blood-letting and cupping. Such methods are likely to have been often debilitating, even if we allow a significant placebo effect.

For most acute diseases – infections, heart problems, dangerous congenital diseases, cancer among the aged – there was little effective response. As the historian Geoffrey Lloyd has observed, in that classic work of Hippocratic observation the *Epidemics*, two-thirds of the patients in the recorded case histories die. Medical authority stemmed less from its ability to cure and more from its practitioners' skill in predicting, explaining and sometimes mitigating illness. That this authority was far from unquestioned is demonstrated by the fact that several Hippocratic texts explicitly engage with criticisms of physicians and the 'medical art' (*iatrike techne*). *On the Art* explains instead that patients do not call in doctors early enough and that surviving illness depends on the relative strength of patient and disease. Failure is due to weakness of the patient or strength of the disease: it is not due, he insists, to medical incompetence.

Some centuries later, Galen and others employed a more subtle defence, relying on the complex interplay of inherited, environmental, individual and lifestyle factors that formed the Greco-Roman account of physiology and illness. Age, gender, regional climate, local meteorology, foodstuffs, daily habits and inheritance created an equation that could only be decoded in probabilistic terms. An individual's sensitivity to particular conditions was determined partly by what they had eaten three weeks before. In this difficult epistemological climate, medicine was necessarily a stochastic enterprise. The commonplace occurrence of morbidity and death, and the inability of doctors to reliably counter either, was a mark not of incompetence but of the difficulty of the enterprise, the inevitability of disease. This is the conceptual structure behind the classical world's concern to treat the patient as a whole, holistically: to take all potential factors into consideration.

The Hippocrates of Cos who appears in the medical literature of today is a successful invention but a decontextualised fragment of history. An image of medicine coalesced around him in antiquity, perhaps precisely because the range of opinions, subjects and attitudes visible in the texts attributed to this shadowy figure enabled authors like Galen to partially create him as a flexible authority permanently subject to interpretation. He came to represent not only all the Hippocratic texts but the entire tradition of Greco-Roman medicine, a historical compression aided by Galen's repackaging of the man and the corpus within his own comprehensive and overwhelmingly influential output. The further he receded in time, the more authoritative Hippocrates became as an ideal of the medicine whose image he was being created in: 'rational', scientific, ethical. He became a culture hero: the 'father of medicine' and a personalisation of scientific authority. It is, as we have seen, problematic as a description of ancient medicine in its original socio-cultural context and its status as a forerunner of modern biomedicine can be made with hindsight only.

The detached, floating nature of Hippocratic aphorisms, almost entirely separated from their cultural or even textual context, enables modern writers, patients and physicians to read our concerns into the headline messages of the ancient corpus. This is often a lazy rhetorical move that does nothing for either history or science; instead the appeal to ancient authority and shorthand ethics obscures or even replaces what our own argument is actually about. Hippocrates should represent our understanding of where medicine has historically come from or how social pressures mould medicine, but more often than not all he indicates is a vague aspiration of what we would like medicine to be. It is one ancient legacy that modern science could do without.

SOME SUGGESTIONS FOR FURTHER READING

GENERAL

A concise but surprisingly comprehensive guide to classical period biology, astronomy and physics, which is holding up well to the passage of time, can be found in G. E. R. Lloyd, *Early Greek Science* (London, 1970) and *Greek Science after Aristotle* (London, 1973). More in-depth analysis can be found in the same author's more academic works, which cover a wide range of subjects and themes in ancient science. These include *Magic, Reason and Experience* (Cambridge, 1979), *Science, Folklore and Ideology* (Cambridge, 1983), *The Revolutions of Wisdom* (Cambridge, 1987) and *Adversaries and Authorities: Investigations into Ancient Greek and Chinese Science* (Cambridge, 1996). Another excellent and original general guide is T. E. Rihll, *Greek Science* (Oxford, 1999).

CHAPTER I

There are several translations available of selected surviving quotations and paraphrases (known collectively as 'fragments') of the Pre-Socratic philosophers, including G. S. Kirk, J. E. Raven and M. Schofield, *The Presocratic Philosophers* (Cambridge, 2nd edn 1983, reprinted 1995); or Patricia Curd and Richard D. McKirahan, *A Presocratics Reader* (Indianapolis, 2nd edn 2011). The Pre-Socratics are a well-discussed topic: a stimulating introduction is Catherine Osborne, *The Presocratics:*

A Very Short Introduction (Oxford, 2004) or at more traditional length, James Warren, *Presocratics: Natural Philosophers Before Socrates* (Stocksfield, 2007). For the committed, Jonathan Barnes, *The Presocratic Philosphers* (London, 1979) offers an in-depth high-powered philosophical analysis of Pre-Socratic thought.

The relationship between the Pre-Socratics and the origins of science and philosophy in Greece is explored by G. E. R. Lloyd, 'The social background of early Greek philosophy and science', republished in Lloyd, *Methods and Problems in Greek Science: Selected Papers* (Cambridge, 1991). See also his *Adversaries and Authorities* (Cambridge, 1996). For influential accounts of the physical nature of the universe and its origins by the early atomists, Plato, and Aristotle, see D. S. Furley, *The Greek Cosmologists Volume 1* (Cambridge, 1987) and David Sedley, *Creationism and its Critics in Antiquity* (Berkeley, 2007).

There are numerous good popular accounts of modern cosmology, including the discovery of cosmic microwave background radiation. Simon Singh, *The Big Bang: The Most Important Scientific Discovery of All Time And Why You Need to Know About It* (London, 2004) starts with the ancient Greeks; Marcus Chown, *Afterglow of Creation: Decoding the Message from the Beginning of Time* (London, revised edition 2010) also concentrates on the big bang. A more general look at the unsolved problems is by string theory expert Brian Greene, *The Fabric of the Cosmos: Space, Time and the Texture of Reality* (London, 2004). The shortest exploration of the topics available is probably Peter Coles, *Cosmology: A Very Short Introduction* (Oxford, 2001).

CHAPTER II

In regard to the two Pre-Socratics discussed at greatest length in this chapter, see Brad Inwood, *The Poem of Empedocles: A Text and Translation* (Toronto, revised edition 2001). This includes both new material and a lengthy introduction. For Anaximander, a recent volume puts together three studies: Dirk L. Couprie, Robert Hahn and Gerard Naddaf, *Anaximander in Context* (Albany, 2003). An individualistic take from a modern perspective is C. Rovelli, *The First Scientist: Anaximander and his Legacy* (translated edition Yardley, 2011).

On the thought and methods of Charles Darwin, the best biography available is Janet Browne's comprehensive and considered two volumes: *Charles Darwin: A Biography* (two-volume edition London, 2003), making full use of her work in F. Burkhardt et al. (eds), *The Correspondence of Charles Darwin* (Cambridge, 1985–). Published in chronological order these so far cover the years up to 1873 (vol. 21, 2014). Two classics of popular science writing on the subject of evolution by natural selection are Richard Dawkins, *The Blind Watchmaker* (Harlow, 1986; new edition London, 2006) and Jonathan Weiner, *The Beak of the Finch: A Story of Evolution in Our Time* (London, 1994).

CHAPTER III

Aristotle's *History of Animals* and *Generation of Animals* are translated as part of Jonathan Barnes, *The Complete Works of Aristotle* (Oxford, 1984). There is an accessible summary of his work in G. E. R. Lloyd, *Early Greek Science* (London, 1970). A general account of the man and his work is J. L. Ackrill, *Aristotle the Philosopher* (Oxford, 1981); for a compressed but stimulating version see Jonathan Barnes, *Aristotle: A Very Short Introduction* (Oxford, 2000); as well as Andrea Falcon, *Aristotle and the Science of Nature: Unity Without Uniformity* (Cambridge, 2005) and the articles collected by Allan Gotthelf and James Lennox (eds), *Philosophical Issues in Aristotle's Biology* (Cambridge, 1987). For natural history in antiquity overall, see Roger French, *Ancient Natural History* (London, 1994).

Philoponus' *Commentary on Aristotle's Physics* is available in translations by A. R. Lacey, *Philoponus: On Aristotle's Physics 2* (London, 1993) and M. Edwards, *Philoponus: On Aristotle's Physics 3* (London, 1994). For discussion of his work, its criticisms of Aristotle and its relationship to later theories of inertia, see Richard Sorabji, *Matter, Space and Motion: Theories in Antiquity and their Sequel* (London, 1988) and the collection of articles by various authors in Sorabji (ed.), *Philoponus and the Rejection of Aristotelianism* (London, 1987).

For Galileo see Stillman Drake, *Galileo at Work: His Scientific Biography* (Chicago, 1978) and Peter Machamer (ed.), *The Cambridge Companion to Galileo* (Cambridge, 1998). There is a voluminous bibliography on Isaac Newton. R. S. Westfall, *Never At Rest: A Biography of Isaac Newton* (Cambridge, 1980) offers a thorough biography; a more abridged version by the same author is *The Life of Isaac Newton* (Cambridge, 1993). J. W. Herivel, *The Background to Newton's Principia* (Oxford, 1965) covers Newton's investigations into dynamics.

For Galen and Erasistratus, see bibliography for Chapter IV.

CHAPTER IV

Helen King, *Greek and Roman Medicine* (Bristol, 2001) offers a taster of ancient medicine. Wesley D. Smith, *The Hippocratic Tradition* (Ithaca, 1979) is the most comprehensive rebuttal of the idea that we can reliably identify Hippocrates of Cos with specific medical treatises; the essentials of the argument can also be found in G. E. R. Lloyd, 'The Hippocratic Question' in Lloyd, *Methods and Problems* (Cambridge, 1991). Good general works on Roman medicine are Ralph Jackson, *Doctors and Diseases in Ancient Rome* (London, 1998) and John Scarborough, *Roman Medicine* (Ithaca, 1969).

Influential works on the concepts of medical anthropology, including the distinction between health and disease, are those of Arthur Kleinman, 'Concepts and a model for the comparison of medical systems as cultural systems', *Social Science and Medicine* 12.2B (1978), pp. 85–93 and *Patients and Healers in the Context of Culture: An Exploration of the Borderland Between Anthropology, Medicine, and Psychiatry* (Berkeley,

1980). Mirko D. Grmek, *Diseases in the Ancient Greek World* (Baltimore, 1989), Robert Sallares, *The Ecology of the Ancient Greek World* (Ithaca, 1991) and ibid., *Malaria in Rome: A History of Malaria in Ancient Italy* (Oxford, 2002) are detailed and comprehensive works on the epidemiology of the ancient world.

There are many translations of the ancient medical texts themselves, although some works remain untranslated into English or exist only in expensive academic editions. G. E. R. Lloyd (ed.), *Hippocratic Writings* (Harmondsworth, 1978) offers a good selection of works from the 'Hippocratic' corpus and some of Galen's shorter treatises can be found in Peter Singer (ed.), *Galen: Selected Works* (Oxford, 1997). A single lengthy text by Soranus on gynaecology is translated by Oswei Temkin, *Soranus' Gynecology* (Baltimore, 1956; republished with additional material 1991). Extracts from rarer authors and works can be found in anthologies: James Longrigg, *Greek Rational Medicine: Philosophy and Medicine from Alcmaeon to the Alexandrians* (London, 1993) is a good selection, although Longrigg's positivist and anachronistic portrayal of ancient medicine as the progressive triumph of rationality over superstition is an outlier in academic discourse. It includes hard-to-find excerpts from the Alexandrian physicians Herophilus and Erasistratus; for which see also the medical section in Georgia Irby-Massie and Paul Keyser (eds), *Greek Science of the Hellenistic Era: A Sourcebook* (London, 2002). The definitive edition and analysis of the fragmentary works of Herophilus is by Heinrich von Staden, *Herophilus: The Art of Medicine in Early Alexandria* (Cambridge, 1989). There is an extensive bibliography on Galen: an up-to-date recent work concentrating on Galen's self-presentation and practice is Susan Mattern, *Galen and the Rhetoric of Healing* (Baltimore, 2008). The pioneering work of William Harvey and its social and intellectual contexts is examined by Thomas Wright, *William Harvey: A Life in Circulation* (Oxford, 2012).

The gynaecological texts of the Hippocratic corpus are particularly difficult to find in translation, although *The Seed* and *The Nature of the Child* can be found in Lloyd (ed.), *Hippocratic Writings*. Important excerpts, together with selections from Aristotle, Soranus, Galen and others are included in the medical section of the anthology by Mary Lefkowitz and Maureen Fant (eds), *Women's Life in Greece and Rome: A Sourcebook* (Baltimore, 1982). An abridged version of this anthology is online at Ross Staife, *Diotima: Materials for the Study of Women and Gender in the Ancient World*. Available at www.stoa.org/diotima (accessed 2 February 2015). The site also contains a comprehensive bibliography on gynaecology, sexuality and related subjects. Among academic works on women and medicine in antiquity are Lesley Dean-Jones, *Women's Bodies in Classical Greek Science* (Oxford, 1994); Ann Ellis Hanson, 'The medical writers' woman', in David M. Halperin, John J. Winkler and Froma I. Zeitlin (eds), *Before Sexuality: The Construction of Erotic Experience in the Ancient Greek World* (Princeton, 1990); and G. E. R. Lloyd, *Science, Folklore and Ideology* (Cambridge, 1983). For the arterial experiment, see Bryan Mowry, 'From Galen's theory to William Harvey's theory: a case study in the rationality of scientific theory change', *Studies in History and the Philosophy of Science* 16 (1985), pp. 49–82.

A good translation of Thucydides' *The Peloponnesian War*, with much supplementary material, is Robert B. Strassler (ed.), *The Landmark Thucydides: A Comprehensive Guide to the Peloponnesian War* (London, 1996). The role of medical thought in the work of the historian Herodotus and fifth-century Athenian culture is analysed by Rosalind Thomas, *Herodotus in Context: Ethnography, Science and the Art of Persuasion* (Cambridge, 2000).

On germ theories of the nineteenth century and the work of Robert Koch and Louis Pasteur, see Agnes Ullmann, 'Pasteur-Koch: Distinctive ways of thinking about infectious diseases', *Microbe* 2.8 (August 2007), pp. 383–7, and John Waller, *The Discovery of the Germ: Twenty-five Years that Transformed the Way We Think About Disease* (London, 2002). A vivid and wide-ranging account of John Snow's identification of the role of the Broad Street pump in a London cholera epic is told by Steven Johnson, *The Ghost Map: The Story of London's Most Terrifying Epidemic – and How It Changed Science, Cities and the Modern World* (London,

2006). For the pioneering work on hospital hygiene by Ignaz Semmelweis, see S. Nuland, *The Doctor's Plague: Germs, Childbed Fever and the Strange Story of Ignaz Semmelweiss* (London, 2004).

CHAPTER V

Serafina Cuomo, *Ancient Mathematics* (London, 2001) is an excellent introduction to the topic, placing mathematics within Greco-Roman social and cultural history. Reviel Netz, *The Transformation of Early Mediterranean Mathematics: From Problems to Equations* (Cambridge, 2004) concentrates on the nature of the mathematics itself, as in his new and much-needed translations of important texts by Archimedes: Netz, *The Works of Archimedes* (Cambridge, 2004–). Ian Mueller, 'Greek mathematics to the time of Euclid', in Mary Louise Gill and Pierre Pellegrin (eds), *A Companion to Ancient Philosophy* (Oxford, 2006), pp. 686–718 offers a guide to early Greek mathematics, including Archytas of Tarentum. For more specialised purposes Carl Huffman, *Archytas of Tarentum* (Cambridge, 2005) is unlikely to be superseded for some time. Ancient astronomy is integrated with modern knowledge in James Evans, *The History and Practice of Ancient Astronomy* (Oxford, 1998).

There is still relatively little work on technological culture, but several recent works go a considerable way towards rectifying the lack. In English these include Serafina Cuomo, *Technology and Culture in the Ancient World* (Cambridge, 2007) and John P. Oleson (ed.), *The Oxford Handbook of Engineering and Technology in the Classical World* (Oxford, 2007). The primary sources are accessible through John W. Humphrey, John P. Oleson and Andrew N. Sherwood (eds), *Greek and Roman Technology: A Sourcebook* (London, 1997) and translations of the ballistic works of Philo and Hero by E. W. Marsden, *Greek and Roman Artillery* (2 volumes, Oxford, 1969). Tracey Rihll, *The Catapult: A History* (Yardley, 2007), by an expert on ancient mechanics and its relationship to society, does what it says on the tin. An older work by J. G. Landels, *Engineering in the Ancient*

World (Berkeley, revised edition, 2000) remains valuable. For Rome's water supply infrastructure see also A. Trevor Hodge, *Roman Aqueducts and Water Supply* (London, 2002). Frontinus' original report is translated by C. E. Bennett and M. B. McElwain, *Frontinus: Stratagems. Aqueducts of Rome*, Loeb Classical Library no. 174 (Cambridge, Mass., 1925). The Roman architect Vitruvius' treatise, a vital source for ancient technology, is published in English as Ingrid D. Rowland and Thomas Noble Howe (eds), *Vitruvius: Ten Books on Architecture* (Cambridge, 2001). Sylvia Berryman, *The Mechanical Hypothesis in Ancient Greek Natural Philosophy* (Cambridge, 2009) is a stimulating corrective to the idea that these two fields were always at odds.

CHAPTER VI

Celsus' medical encyclopaedia is translated in three volumes by W. G. Spencer, *Celsus: On Medicine*, Loeb Classical Library nos 292, 304 and 336 (Cambridge, Mass., 1935–8). For ancient attitudes to animals, see Richard Sorabji, *Animal Minds and Human Morals: The Origins of the Western Debate* (Ithaca, 1993).

An excellent introduction to the history and practice of astrology is Tamsyn Barton, *Ancient Astrology* (London, 1994). At greater length, both historically and literally, is S. J. Tester, *A History of Western Astrology* (Woodbridge, 1987). Liba Taub, *Ptolemy's Universe: The Natural, Philosophical and Ethical Foundations of Ptolemy's Astronomy* (Chicago, 1993) gives an overview of the most influential astronomer–astrologer of antiquity. Frank E. Robbins, *Ptolemy Tetrabiblos*, Loeb Classical Library no. 435 (Cambridge, Mass., 1940) is still the English translation of his astrological treatise, although there is a newer rendition of his principal work on astronomy: G. J. Toomer, *Ptolemy's Almagest* (Princeton, 1998).

For the earlier use of Hippocrates in fact, fiction, and mixtures of the two, see Jody Rubin Pinault, *Hippocratic Lives and Legends* (Leiden, 1992) and Wesley D. Smith, *The Hippocratic Tradition* (Ithaca, 1979).

NOTES

Chapter I

1 The surviving evidence of the Pre-Socratics' lives, works and ideas is translated and commented upon in several modern collections. I have normally used G. S. Kirk, J. E. Raven and Malcolm Schofield (eds), *The Pre-Socratic Philosophers* (Cambridge, 2nd edn, 1983).
2 It might be thought that the atomists' intellectual descendants, the followers of Epicurus, argued that reality was directly apprehensible; but in fact such perceptions, while true, offered only a partial picture. An observer may correctly perceive a tower, at different times, as both large (close to) and small (a long way off) but it is the fact that both these perceptions are true that needs explaining. See Anthony A. Long and David N. Sedley, *The Hellenistic Philosophers*, Vol. 1 (Cambridge, 1987).
3 Hesiod's 'Chaos' shares conceptual terrain and probably partly derives from other Near Eastern creation myths (including those of Egypt, Mesopotamia and the Levant) which involve a creative act that reforms some primeval substance – in Egypt's case the creative waters of the goddess Nun – into the ordered world. These elements were reformulated in Pre-Socratic cosmogonic and cosmological speculation, as in Thales' proposal that the earth floated on water and (if he did in fact say this) that water was the original element from which the others derive.
4 For this and alternative theories see e.g. Peter Coles, *Cosmology: A Very Short Introduction* (Oxford, 2001) p. 121.
5 For the pervasiveness of this concept in Greco-Roman thought see the references collected by David Sedley, *Creationism and its Critics in Antiquity* (Berkeley, 2007), p. 146 n. 29.
6 For the different kinds of monism and which kind Parmenides might have been espousing, see John Palmer, 'Parmenides', in Edward N. Zalta (ed.), *The Stanford Encyclopedia of Philosophy*. Available at http://plato.stanford.edu/

archives/sum2012/entries/parmenides. (Summer 2012 edition, accessed 28 January 2015).

7 The astronomer Edwin Hubble observed redshift (as the source of light moves away, the crests of the light radiation emanating from it become further apart as its wavelength lengthens), which indicated that stars were moving away from earth at a speed proportional to their distance from earth, demonstrating that the universe was expanding.

8 On the Greek concept 'nature' or *physis*, developed between the sixth and fourth centuries BC, and on Aristotle's biological investigations, see Chapter III.

9 The atomism of Leucippus and Democritus was a minority view among Greco-Roman intellectuals, but was nonetheless a notorious and indirectly influential theory. Opponents needed to develop arguments against it and to produce persuasive explanations of their own about physics, perception and determinism. Much of what we know about early atomism is derived from the discussion by Aristotle, who disagreed with atomism in almost every particular. In the third century BC Epicurus of Samos and his followers, the Epicureans, adopted and adapted atomic theory, developing or altering Democritus' views.

10 According to Plato in his *Theaetetus*, 152b1–c8, who discusses the idea in detail, Protagoras said, 'Man is the measure of all things: of the things that are, that they are; and of the things that are not, that they are not.'

11 Kirk, Raven and Schofield, *Pre-Socratics*, Fragment 565.

12 See below as to how this inevitability of unlikely things, so long as they are atomically possible, might be relevant to biology and evolution.

13 Exactly what Epicurus and his followers thought the gods to be is obscure and controversial. The notion of an eternal and perfect being should conflict with the Epicurean doctrine that only atoms – not atomic structures – last forever. Two different intepretations are argued by, respectively, David Sedley and David Konstan, in K. Sanders and J. Fish (eds), *Epicurus and the Epicurean Tradition* (Cambridge, 2011), pp. 29–52 and 53–71.

14 The former are now nominatively associated with hedonism and the latter with endurance of pain: these are both garbled and compressed versions of their originators' actual doctrines.

15 For language of 'before' the cosmos in reference to chaos see e.g. 51d2–53c3: before the creation of the world there was chaotic motion within the receptacle.

16 *Timaeus* 37d5–38a1; 38b7–8.

17 G. Vlastos, 'The disorderly motion in the *Timaeus*', *Classical Quarterly* 33 (1939), pp. 71–83. As Sedley, *Creationism*, 144–5 sets out, this solution also provided an answer to one of the objections to creation itself: why did god choose that moment to launch the cosmos? If there was no passing of finite time before creation, there were no alternative moments to choose between. The moment of creation was the first moment possible.

18 It is also possible to argue, from the mathematical properties of some quantum cosmological theories, that time does not exist in the way it does now in singularity conditions wherein gravity is extreme: on this view the universe did not 'begin', because 'beginning' requires time, and under the conditions in question the universe was literally timeless.
19 There was some doubt over whether all differences would necessarily make an iteration of the universe less perfect than another (which would therefore not happen). Some Stoics argued that very trivial distinctions, such as the placement of a mole on a person's arm, might vary from one iteration to another without affecting anything else. They were unaware of the mathematics of chaos theory and its famous example of a butterfly's wing motion, in which apparently tiny things can have important causal effects.
20 Kirk, Raven and Schofield, *Pre-Socratics*, Fragment 348.

Chapter II

1 G. S. Kirk, J. E. Raven and Malcolm Schofield (eds), *The Pre-Socratic Philosophers* (Cambridge, 2nd edn, 1983), Fragment 134.
2 Ibid., Fragment 135.
3 Ibid., Fragment 134.
4 Ibid., Fragment 91.
5 Letter to J. D. Hooker, February 1871. Published in F. H. Burkhardt et al. (eds), *The Correspondence of Charles Darwin*, Vol. 19 (Cambridge, 2012).
6 Kirk, Raven and Schofield, *Pre-Socratics*, Fragment 184, slightly revised translation.
7 For the reluctance of Lyell to accept the mutability of species, see Janet Browne, *Charles Darwin: A Biography* (London, 2003) especially Vol. 2, *The Power of Place*. Sedgwick's opposition, ibid., p. 336.
8 'Here sprang up many faces without necks, arms wandered without shoulders, unattached, and eyes strayed alone, in need of foreheads.' Kirk, Raven and Schofield, *Pre-Socratics*, Fragment 57 (their translation).
9 Ibid., Fragment 379.
10 Ibid., Fragment 380.
11 Darwin's use of the word 'selection', even when coupled with 'natural', caused problems for the understanding of his theory. In the fifth edition of *Origin* he was persuaded to introduce Herbert Spencer's phrase 'survival of the fittest' but this proved even more easily misunderstood (as representing athleticism with eugenic overtones, rather than the quality of being well adapted to one's environment).
12 M. R. Dietrich, 'Richard Goldschmidt: hopeful monsters and other "heresies"', *Nature Review Genetics* 74.4 (January 2003), pp 68–84, in particular pp. 71–2.
13 Unlike Goldschmidt's other hypothesis for macromutation, in which small chromosomal changes eventually and abruptly led to a drastic change in

phenotype, this idea of developmental macromutation did not require rejecting classical gene theory.

14 David Sedley, *Creationism and its Critics in Antiquity* (Berkeley, 2007), pp. 67–70.

15 (Oxford, 1976). It is of course the case that mutations harmful enough to noticeably impede or kill the individual which carries it do occur, and this is arguably equivalent to a hopeless monster.

16 P. Shipman, *Taking Wing: Archaeopteryx and the Evolution of Bird Flight* (London, 1998).

17 Mendel's work on genetics – inherited characteristics in sweet peas – was contemporary but only published in an obscure Bavarian journal. Darwin was unaware of it. The discovery of the actual structure of DNA by Watson and Crick, utilising the crucial work done by Rosalind Franklin and Maurice Wilkins with X-ray crystallography, would not be made until the mid-twentieth century. James Watson, *The Double Helix* (London, 1968).

18 Charles Darwin, *On the Origin of Species by Means of Natural Selection* (London, 1859), p. 134: 'From the facts alluded to in the first chapter, I think there can be little doubt that use in our domestic animals strengthens and enlarges certain parts, and disuse diminishes them; and that such modifications are inherited.'

19 One means of explaining how such acquired characteristics could be passed on were theories of pangenesis. Darwin's version was similar to early examples formulated by Democritus and authors of the Hippocratic medical corpus. The general idea is that elements within, or fragments of, each part of the body – bone, blood, left shoulder, lips – are passed on from one or both parents. Thus a change to one's right forearm during life, for example scarring from an injury, would form part of the inheritance of any offspring born after that change.

20 Kirk, Raven and Schofield, *Pre-Socratics*, Fragment 386.

21 Browne, *Darwin* Vol. 1: *Voyaging*, p. 339 n. 32. On Darwin and Malthus ibid., pp. 385–90: the quotation is in P. H. Barrett et al. (eds), *Charles Darwin's Notebooks, 1836–44* (Cambridge, 1987), D135, taken from Browne, *Voyaging*, p. 389 n. 43.

22 Quotation from Burkhardt et al., *Correspondence*, Vol. 7 (Cambridge, 1992), p. 127: reference taken from Browne, *The Power of Place*, p. 47.

23 Some of these belonged to a rather different social class, and the information they supplied often became an anonymous accumulation of information, to be combined and utilised by theorists. In a similar manner, in antiquity accounts of fossils such as those used by Xenophanes might have come from low-status and anonymous individuals. Then, too, bee-keepers, farmers, shepherds, fishermen and hunters were exploited as a source of information by empirically minded individuals of a different social-economic status and private income, notably Aristotle and Theophrastus.

24 Notebook M. Barrett et al., *Darwin's Notebooks*, p. 539.

NOTES

Chapter III

1. The medieval theologian Thomas Aquinas developed a theory of moral 'natural law', in which our rational human morality is dependent on our rational human nature, from his readings of Plato and Aristotle. (The phrase is *dikaion physikon* in Greek or *lex naturalis* in Latin.)
2. Robert Mayhew, *The Female in Aristotle's Biology* (Chicago, 2004).
3. Robert M. Pursig, *Zen and the Art of Motorcycle Maintenance: An Inquiry into Values* (London, 1974), p. 98.
4. J. B. S. Haldane, *Possible Worlds and Other Essays* (London, 1927), p. 286.
5. 'There are more things in heaven and earth, Horatio, than are dreamt of in your philosophy.' *Hamlet*, Act I, Scene 5, lines 167–8.
6. The experiment with two holes demonstrates that in fact it does behave like both. Richard P. Feynman, R. Leighton and M. Sands, *The Feynman Lectures on Physics*, Vol. 3 (Reading, Mass., 1965) 1–9. There are numerous popular accounts: for example John Gribbin, *Schrödinger's Kittens and The Search for Reality* (London, 1995).
7. Aristotle, *Physics* 7.8.
8. The story comes from the biography of Galileo by his student Vicenzo Viviani, but Galileo's surviving works do not include mention of him carrying out such a test.
9. Michael Wolff, 'Philoponus and the rise of preclassical dynamics', in Richard Sorabji (ed.), *Philoponus and the Rejection of Aristotelian Science* (London, 1987), pp. 84–120 n. 5.
10. *Physics* 4.8 (215a14–7).
11. Aristotle, *Physics* 4.7 (214a29–32); see also Plato, *Timaeus*, 80c.
12. Wolff, 'Philoponus'.
13. Simplicius, *On Aristotle's Physics*, 1349, 16 (ed. Diels).
14. Ibid., 693, 10 ff. (ed. Diels). 'Gaps' is usually translated as 'pores' or 'ducts' (*poroi*): literally passages, tunnels.
15. Emphasised by Philo of Byzantium, *On the Construction of Catapults* 71.17–72.4. For whether this was a more ad hoc process than a program of systematic experimentation see Tracey Rihll, *The Catapult: A History* (Yardley, 2007; revised edition 2009).
16. The story is anecdotal, but the scientific principles behind the procedure were certainly known at the time.
17. Weight is mass under gravity (W=mg). Dalton's 'atomic mass' would now be called 'relative atomic weight'. Since neither of the periods under discussion distinguished between the two, I have not been consistent either.
18. Lavoisier defined elements as 'All substances that we have not yet been able to decompose by any means.' Modern chemistry defines a 'compound' as deriving from two or more elements that can be decomposed by chemical methods, but not physically separated. Most or all of the properties of a compound will differ

from those of the original elemental substances. A 'mixture', like oil and water, can be physically separated out again.

19 The original discovery of 'dephloginistised air' or oxygen was by Joseph Priestley in 1774, and two years earlier, unpublished, by Carl Scheele Sweden. The story of chemistry in this period as I have summarised it here is a simplified one: the virtues on all sides of the arguments are given fairer treatment by Hasok Chang, *Is Water H₂O: Evidence, Pluralism, and Realism* (Dordrecht, 2012).

20 He was largely right. You can break an atom, it turned out, and even change it, but these changes are nuclear in nature, not chemical.

21 Not counting isotopes.

22 As far as the eighteenth century knew: some complex compounds do not reduce to such simple integer ratios.

23 Although he had to guess at the actual arrangement of atoms in any given compound, he managed to draw up a table of the atomic weight of elements using their ratio by mass in compounds. This was of variable accuracy. He assumed, for example, that water had the simplest possible 'binary' arrangement, so that there was one atom of hydrogen for every one of oxygen, instead of two hydrogen for one oxygen (H_2O). Oxygen is always 8/9 of the mass of any amount of water, with hydrogen making up the remaining 1/9. If hydrogen and oxygen atoms were in a binary ratio of 1:1, then oxygen's atomic weight would be eight times as heavy as hydrogen. In fact, the ratio is 2:1, so in fact oxygen is 16 times as heavy as hydrogen: an atomic weight of 16.

24 Slightly more of it if you count the invisible dark matter out there (probably).

25 Peter Atkins, *Galileo's Finger: The Ten Great Ideas of Science* (Oxford, 2003), p. 146.

Chapter IV

1 Not only theoretical models, but ideas of what constitutes health as opposed to illness, can vary between cultures. Illness is often understood as what varies from socially recognised norms.

2 Cato, *On Agriculture* 160; 156.

3 *Inscriptiones Graecae* IV.2, pp. 121–2, no. 21 = Ludwig and Emma Edelstein, *Asclepius* (Baltimore, 1945), no. 423.

4 Exactly what substance was thought to be black bile is unclear. Very dark bloody matter is a possibility.

5 G. E. R. Lloyd (ed.), *Hippocratic Writings* (Harmondsworth, 1978), pp. 239–40.

6 The author of *On the Sacred Disease* points to the counter-example of Libyan society, which according to him used goat products extensively without suffering universal epileptic seizures: Lloyd, *Hippocratic Writings*, p. 238.

7 E.g. healing spells and practices in the Greek and demotic 'magical' papyri of the late Hellenistic and Roman periods: H. D. Betz (ed.), *The Greek and Demotic Magical Papyri in Translation* (Chicago, 1992).

NOTES

8 The anecdote is from Aulus Gellius, *Attic Nights*, 18.10.
9 This was not a one-way process. Naturalistic physicians used drugs from older traditions, stripping them of incantations and the original reasons for their use, such as sympathetic magic. Dietary regimen is an idea expanded from the folk uses of common substances. The medical text *On Dreams* (= *On Regimen* 4) reinterpreted the notion that gods send advisory or predictive dreams. In its version, dreams are physiologically triggered signs of the body's internal state.
10 For the powers of plant and animal substances in relation to the gods and other para-natural entities see the account of the practices and beliefs of Greece's specialist plant-collectors and drug-preparers, known as the 'root-cutters', by the Aristotelian natural philosopher Theophrastus in his *History of Plants*, 9.8–19.
11 Galen, *On the Temperaments and Properties of Simple Drugs* 6.10 and 10.13 (11.859–60 and 12.207 ed. Kühn).
12 Unusually for antiquity, Galen pursued anatomical investigation repeatedly and carefully.
13 Anon. Lond. 33.44–51. See Heinrich von Staden, 'Experiment and experience in Hellenistic medicine', *Bulletin of the Institute of Classical Studies* 22 (1975), pp. 178–99, especially 179–81.
14 See Chapter III for a similar use of weight before and after a transformative process in a series of experiments by the French chemist Lavoisier, which demonstrated that mass is conserved in any chemical reaction.
15 *On the Natural Faculties* 1.13 (2.646–8 ed. Kühn). Translation by A. J. Brock, *Galen on the Natural Faculties* (London, 1916), pp. 59–62.
16 Galen, *On the Natural Faculties* 3.8 (2.174 ed. Kühn).
17 There would seem to be a simple way to disprove the former – cut an artery, observe blood – but Erasistratean biological physics had a subtle answer to this objection and it was difficult to conclusively disprove. Galen used experiment and argument to attack their defences at length: see Galen, *Whether Blood is Naturally Contained in the Arteries* (=4.703–36 ed. Kühn); translated by David J. Furley and John S. Wilkie, *Galen on Respiration and the Arteries* (Princeton, 1984), pp. 137–83.
18 Ibid., 8.4 (4.733–4 ed. Kühn); Furley and Wilkie, *Galen on Respiration*, pp. 179–81. Cf. Galen, *Anatomical Procedures* 7.16 (2.646–8 ed. Kühn).
19 For the failure of modern repetitions of this experiment see e.g. J. M. Forrester, 'An experiment of Galen's repeated', *Proceedings of the Royal Society of Medicine* 47 (1954), pp. 241–4 and C. R. S. Harris, *The Heart and the Vascular System in Ancient Greek Medicine* (Oxford, 1973), pp. 379–88.
20 E.g. the author of *On the Sacred Disease*, who asserts (in an era before human dissection) that if one cuts open the skulls of goats with a similar disease, the brains will be found to be full of wet phlegm. See also Aristotle's falling bodies and female teeth counting in Chapter III.

21 See Heinrich von Staden, 'Hairesis and heresy: the case of the haireseis iatrikai', in B. F. Meyer and E. P. Saunders, *Jewish and Christian Self-Definition*, Vol. 3 (London, 1982), pp. 76–100, 199–206.
22 Thucydides, *History of the Peloponnesian War*, 2.47.1–55.1.
23 Later Greco-Roman physicians thought of plagues, skin diseases and eye conditions as dangerous in this way. Ordinary fevers, colds and digestive disorders were not usually included, although *On the Nature of Man* attributed fevers to unhealthy exhalations causing bad air.
24 [Hippocrates], *Airs, Waters, Places* 4 (Lloyd, *Hippocratic Writings*, p. 150).
25 Vitruvius 1.4.
26 *On the Affected Parts* 6.5 (8.423–9 ed. Kühn).
27 *On Agriculture* 1.12.2–4.
28 Cf. the second-century AD Roman writer Columella, *On Agriculture* 1.5.6: [From the marshes] [...] issue forth a harmful stench [...] animals with vicious spikes that fly about [...] snakes and vipers filled with deadly venom, from which come unseen diseases that defy doctors to know their causes.' Atomist theory included 'seeds of disease': mobile mechanical disruptions to a functioning body (Lucretius, *On the Nature of Things*, e.g. lines 1093–102).
29 See Vivian Nutton, 'The seeds of disease: an explanation of contagion and infection from Greeks to the Renaissance', *Medical History* 27 (1983), pp. 1–34; reprinted in ibid., *From Democedes to Harvey* (London, 1988), Chapter XI.
30 Aristotle, *History of Animals* 9.1 (608a33–b18).
31 Xenophon, *Oeconomicus* 10.1; 7.20–8.
32 'Further, a boy actually resembles a woman in physique, and a woman is as it were an infertile male; the female, in fact, is female on account of inability of a sort, viz. it lacks the power to concoct semen out of the final state of the nourishment (this is either blood, or its counterpart in bloodless animals) because of the coldness of its nature.' Aristotle, *Generation of Animals*, 1.17 (716a). It is unclear what Aristotle would have thought of the modern embryological discovery that the human foetus defaults to female. A male reproductive anatomy only develops if a gene on the Y chromosome triggers a testis-determining specific factor.
33 Most texts on the subject assume that women produce semen, albeit in less noticeable amounts than the man and discharged directly into the womb (i.e. not visible externally).
34 [Hippocrates], *On Regimen*, 1.18–9.
35 See further Lesley Dean-Jones, 'The politics of pleasure: female sexual appetite in the Hippocratic Corpus', *Helios* 19 (1992), pp. 72–91.
36 Soranus, *Gynaecology* 1.37.
37 [Hippocrates], *On the Nature of the Child*, 5.
38 Aristotle investigated embryology by taking several fertilised chicken eggs and opening one each day, a process which enabled him to observe the course of fetal development. Aristotle, *History of Animals* 6.3 (561a4–21).

Chapter V

1. Reviel Netz, 'Classical mathematics in the classical Mediterranean', *Mediterranean Historical Review* 12 (1997), pp. 1–24; see also ibid., 'Greek mathematicians: a group picture', in C. J. Tuplin and T. E. Rihll, *Science and Mathematics in Ancient Greek Culture* (Oxford, 2002), pp. 196–216, to which much of the following discussion is indebted.
2. None of these three were natives of Alexandria, a centre of scientific activity under the patronage of the early Ptolemies. Conon was from the Greek island city of Samos and Eratosthenes from Cyrene in Greek Libya. Dositheus was from Egypt: the northeastern Delta.
3. Not to mention his inaccurate application of Pythagoras' theorem to the length of ladders used in storming a town: Netz, 'Group picture', 212–3.
4. The actual royal road of the time had been built by the Persian kings for their couriers, and was over 1,600 miles long.
5. In the northern hemisphere. This is now known as the 'tropical' year, in which the sun moves 360° longitudinally in the plane of the ecliptic (the path of the sun through the heavens as seen from earth). It is about 20 minutes less than the 'sidereal' year – the time taken for the sun to return to the same point against the background of the non-planetary stars.
6. Agathemerus, fragments 12 A 6 and 68 B 15 (ed. Diels-Kranz).
7. According to the third-century BC geographer and mathematician Eratosthenes of Cyrene, as reported by Strabo, *Geography*, 1.1.
8. Kurt A. Raaflaub and Richard J. A. Talbert, *Geography and Ethnography: Perceptions of the World in Pre-Modern Societies* (Chichester, 2010), p. 147.
9. See further Christian Jacob, *The Sovereign Map: Theoretical Approaches in Cartography through History* (Chicago, 2006).
10. During the Apollo 8 lunar mission in 1968 astronaut William Anders was having difficulty identifying features on the earth's surface. He realised suddenly that the southern hemisphere was 'on top' – his orientation had reversed in the weightless conditions of the command module. Andrew Chaikin, *A Man on the Moon: The Voyages of the Apollo Astronauts* (London, 1994), p. 96.
11. James Evans, *The History and Practice of Ancient Astronomy* (Oxford, 1998), p. 102.
12. For attitudes on this subject from the fifth and fourth centuries BC, see Herodotus, *Histories*, 4.36 and Aristotle, *Meteorology*, 2.5 (362b12).
13. For a classic statement of geography's relation to astronomy, and what social classes of people are concerned with this kind of knowledge, see Strabo, *Geography*, 2.5.1.
14. Strabo, *Geography*, 1.1.20; Aristotle, *On the Heavens*, 2.14 (297b23–298a121).
15. Ptolemy, *Almagest*, 3.1.
16. Evans, *History and Practice*, 99–100. Vintager is a star in Virgo: its morning rising marked the beginning of the grape harvest. The *parapegmata* are a good

illustration of how Greek society regarded astronomy and meteorology as inextricably linked.
17 Ibid., 63–7.
18 Eratosthenes was aware that he could not be precisely sure of all his information. He assumed that Alexandria and Syene are on the same north–south arc (meridian): in fact there is a 2° difference. Syene is not quite at the latitude where the sun is directly overhead at zenith (the Tropic of Cancer). The distance between the cities was an estimate, and the earth is not a perfect sphere.
19 Vitruvius, *On Architecture*, 1.1.4.
20 *On Architecture* is one of our principal sources for Vitruvius' technical ancestors the Greek mechanists, in particular for the work of Ctesibius and Hero (both of Alexandria).
21 Frontinus, *On the Aqueducts of the City of Rome*.
22 *Ancient Mathematics* (London, 2001), p. 155.
23 The Mars Polar Lander was lost due to unstandardised metric and imperial measurements in its programming; similar problems afflicted the original Hubble telescope and contributed to the command module explosion on Apollo 13.
24 E.g. the inscription of the 'Res Gestae' ('Things Done'), which memorialises the emperor Augustus' achievements. Resources, citizens, colonies and battles are described almost entirely in terms of numbers.
25 The English word 'device' has a similar semantic range, but is not now widely used in the former sense.
26 Later – much later – accounts describe Archimedes as aiding in the defence of Syracuse from the beseiging Roman fleet by inventing war machines the Claw and the Burning Mirrors. A good place to start is Chris Rorres, *Archimedes* (New York University). Available at http://www.math.nyu.edu/~crorres/Archimedes/contents.html. Accessed 4 February 2015.
27 Diogenes Laertius, *Lives of the Philosophers*, 8.83; see discussion by Sylvia Berryman, *The Mechanical Hypothesis in Ancient Greek Natural Philosophy* (Cambridge, 2009) 87–97; Carl A. Huffman, *Archytas of Tarentum: Pythagorean, Philosopher and Mathematician King* (Cambridge, 2005), e.g. p. 14.
28 See Huffman, *Archytas*.
29 From Eutocius' commentary on Archimedes' *On the Sphere and Cylinder* (2.88.3–96.1 ed. Heiberg).
30 Tracey Rihll, *The Catapult: A History* (Yardley, 2007).
31 Plutarch, *Life of Marcellus*, 14.5–6.
32 For armillary spheres see Evans, *History and Practice*, Chapter Two. A useful online source including images is 'The Armillary Sphere' at the Whipple Museum for the History of Science, Cambridge University. Available at http://www.hps.cam.ac.uk/starry/armillary.html (accessed 4 February 2015).

33 Moreover, since the linkage had an odd number of meshings, the two gears would have turned in opposite directions, so there was an identical 'contrate' gear above and linked to the sun gear, which turned at the same speed but in the opposite (right) direction.
34 Y. Freeth, Y. Bitsakis, X. Moussas et al., 'Decoding the ancient Greek astronomical calculator known as the Antikythera mechanism', *Nature* 444 (30 November 2006), pp. 587–91. See also J. V. Field and M. T. Wright, 'Gears from the Byzantines: a portable sundial with calendrical gearing', *Annals of Science* 42 (1985), pp. 87–138.
35 Alan M. Turing. 'On computable numbers, with an application to the Entscheidungsproblem', *Proceedings of the London Mathematical Society* 42.2 (1936).
36 John Tzetzes, *Chiliades* 2.129–30.
37 He is said to have used this quotation in an interview only a few hours after the Los Alamos test. For references, variants and implications see James A. Hijaya, 'The *Gita* of J. Robert Oppenheimer', *Proceedings of the American Philosophical Society* 144.2 (June 2000), pp. 122–67.
38 Pliny, for instance, criticised purple dye, extracted from the shellfish of the Tyrian coast to colour the Roman emperor's clothes, as an unnatural invention of luxury and moral laxity: society could not let nature well enough alone. *Natural History*, 9.62.135 ff.
39 Fragments 1–3 in Huffman, *Archytas*.
40 Steven J. Brams and Alan D. Taylor, 'An envy-free cake division protocol', *The American Mathematical Monthly* 102.1 (1995), pp. 9–18; Ian Stewart, 'Fair shares for all', *New Scientist* 1982 (17 June 1995). Available at http://www.newscientist.com (accessed 18 January 2014). See also Brams and Taylor, *Fair Division: From Cake-Cutting to Dispute Resolution* (Cambridge, 1996) and most recently Steven Brams, Michael A. Jones and Christian Klamler, 'Better ways to cut a cake', *Notices of the American Mathematical Society* 53.11 (2006), cols 1314–421.
41 Berryman, *Mechanical Hypothesis*, pp. 201–15.
42 Translation (slightly adapted) from E. W. Marsden, *Greek and Roman Artillery: Technical Treatises* (Oxford, 1971), pp. 1–2.
43 Compared to the Stoics, Voltaire's Dr Pangloss was a rank amateur.
44 For a guide to the major philosophical schools founded in the Hellenistic period, see e.g. A. A. Long, *Hellenistic Philosophy: Stoics, Epicureans, Sceptics* (London, 1974) and the collection and discussion of relevant fragments by A. A. Long and David N. Sedley, *The Hellenistic Philosophers* (Cambridge, 1987).
45 They employed or bought people to do the actual labour.
46 See e.g. *Topics*, 145a15–16; *Physics*, 192b8–12; *On the Heavens*, 298a27–32, *On the Soul*, 403a27–b2; *Metaphysics*, 1025b25, 1026a18–9, 1064a16–9, b1–3; *Nicomachean Ethics*, 1139a26–8, 1141b29–32.

47 E.g. Simon Blackburn, *Think!* (Oxford, 1999).
48 Personal informal communications with various scientists. I am caricaturing the debate, but not out of recognition.

Chapter VI

1 Another piece of evidence, from the second-century AD Christian writer Tertullian, corroborates it as a contemporary belief.
2 Galen, *Anatomical Procedures*, 9.11.
3 One study on animal trials in relation to stroke found that only one-third of results in animals were replicated in humans, due to publication bias: M. R. Macleod et al., 'Publication bias in reports of animal strokes leads to major overstatement of efficacy', *PLoS Biology* 8.3 (2010), e1000344.
4 Steven Rose, *The Making of Memory* (London, 1993).
5 See the by now much cited psychological experiment on the neuroscience of the 'trolley problem', originally an ethical dilemma by the philosopher Philippa Foot. J. D. Greene, 'The secret joke of Kant's soul', in W. Sinnott-Armstrong (ed.), *Moral Psychology* Vol. 3: *The Neuroscience of Morality* (Cambridge, Mass., 2008).
6 A stronger theory of the physics of astrology involved the concept of 'sympatheia': which, very roughly, says that all things in the universe are connected to each other; sometimes in ways that have no obvious causal transmission but nonetheless correlate predictively. In Stoic philosophy, for example, the flight of birds correlates with human events, enabling accurate divination. This is a long shot, but sympathy-theorists might have approved of the quantum phenomenon which Einstein called 'spooky action at a distance': when one of an entangled pair of photons is observed, the probability waveform of the other also collapses into a determined state, however far apart they are. The speed of light is not a restriction.
7 There is also these days a thirteenth zodiac constellation, Ophiuchus.
8 Plato, *Protagoras* 311b–c, *Phaedrus* 270c–d; Aristotle, *Politics* 4.4. Some kind of corpus was established in the Library of Alexandria by c.250 BC: Vivian Nutton, *Ancient Medicine* (London, 2004), p. 61.
9 E.g. by Galen, who took *On the Nature of Man*, the four-humours text, to be central to the canon and excluded those 'Hippocratic' texts that seemed the most divergent from its language and opinions.
10 Plato's remarks in the *Phaedrus* are not very clear and seem rather Platonic, suggesting interpretation rather than report. Aristotle mentions Hippocrates only as a famous physician, without citing any theories or treatments. A doxography from an Aristotelian context (Anon. Lond. 5.35–6, 44) ascribes to Hippocrates an (interpreted) set of views out of line with humoural theory in many corpus texts and therefore later 'Hippocratic' tradition. They have similarities with *Breaths*, a Hippocratic treatise usually thought simplistic and inferior.

NOTES

11 E.g. *On the Heart*, which clearly post-dates third-century BC anatomical discoveries.
12 For the biographical traditions around Hippocrates' name, see Jody Rubin Pinault, *Hippocratic Lives and Legends* (Leiden, 1992). Many of these stories were also linked to other famous physicians.
13 In scientific journalism, based on my unscientific survey of the three publications mentioned.
14 *The Economist* (28 August 2008).
15 'The good, the fad and the unhealthy', *New Scientist* (23 September 2006).
16 *New York Times* (14 May 2011).
17 'Fever: friend or foe?', *New Scientist* (5 August 2010); see also *Nature Immunology* (DOI: 10.1038/ni1406).
18 D. Kwiatkowksi and M. Nowak, 'Periodic and chaotic host-parasite interactions in human malaria', *Proceedings of the National Academy of Sciences of the USA* 88 (June 1991), cols 5111–13.

INDEX

Agathemerus 141
Airs, Waters, Places 112
Alexander the Great 95–6, 173
Alexandria 96, 98, 124, 131, 138, 144, 146–7, 171, 175, 177–8, 188, 219n2, 220n18
Anaxagoras 4, 25, 183
Anaximander 1, 8–9, 11–24, 28–31, 36, 140–2, 144
Anaximenes 1, 4
Anonymus Londinensis papyrus 99
Antikythera device 158–61, 163
antiperistasis 69–70, 76
Aratus 132
Archimedes 3, 80, 131–2, 146, 153, 162, 220n26
Archytas 154–7, 164–5, 167
Aristarchus 146
(Aelius) Aristides 91
Aristophanes 55–6, 122, 135–6
Aristotle 2–4, 12–15, 17, 19–20, 24, 30–3, 36, 40–1, 50, 52–3, 57–73, 75–6, 87, 99, 103, 117–20, 126–7, 130, 144–5, 152–4, 171, 179, 188, 208, 212n9, 214n23, 215n1, 217n20, 218nn32, 38; 219n12, 222n10
Art, On the 201
Asclepiades 101–3, 109
Asclepius, temples of 90–1, 95–6
ataraxia 168–70
Athens 55, 109, 115, 135–6, 145
Augustus 148–9, 184, 220n24
Avicenna (Ibn Sīnā) 67, 73

Babylon, Babylonia 1, 95, 129, 133, 141, 150
Babylonian astronomy/astrology; *see also* Chaldeans 95, 146–7, 182–4, 187
Beck, Harry 144
Berryman, Sylvia 167
big bang, the 5–7, 9, 11–12, 20–1, 23–5, 27

Callippus 159
Calvisius Taurus 96
Cato the Elder 90, 95
Celsus 176–9, 200
Chaldeans; *see also* Babylonian astronomy 95, 184–5
Conon 131–2, 219n2
Ctesibius 162, 164, 166, 172

Dalton, John 81–3
Darwin, Charles 33–4, 44–51, 61, 214n17
Delphi 140–1
Democritus 12, 15–19, 24, 37, 76, 83, 85, 120, 212n9, 214n19
Diodorus 173
dissection (general) 60, 62, 98–9, 102, 175–80, 217n20
dissection, human 175–80, 217n20
Dositheus 131, 219n2

Egypt 1, 95, 129, 132–3, 138, 143, 147, 150, 175, 177, 183, 188, 211n3, 219n2
Empedocles 26–7, 31–2, 37–44, 49–50

INDEX

Empiricists (medical sect) 108–9, 179
Epicurus, Epicureans 12, 19–21, 36, 49, 83, 168–9, 211n2, 212n9, 212n13
Erasistratus 58, 92, 98–100, 103–5, 107–9, 166, 175–7
Eratosthenes 131–2, 146–8, 156, 158, 219n2, 219n7, 220n18
Ethiopia 109, 143
Euclid 128, 130, 132
Euctemon 136, 145
Eudoxus 145–6

fossil(s) 34–6, 39, 49, 214n23
Fracastoro, Girolamo 115
Frontinus 149–50

Galen 3, 19–20, 92, 97–98, 101–9, 111, 114, 116, 130, 133, 175–6, 179, 190, 192–3, 200–1, 217nn12, 17; 222n9
Galileo 66–7, 73–5, 215n8
Geminus 145, 158
Goldschmidt, Richard 38

Hadrian 151
Haldane, J. B. S. 63
Harvey, William 106–7
Hecataeus 141
Hero 77–9, 85, 150, 164, 168, 170, 172–3, 220n20
Herodotus 56, 110, 143
Herophilus 58, 104, 107–9, 124, 166, 175–7, 190
Hesiod 8
Hipparchus 187
Hippocrates 92, 188–93, 195–7, 201–2, 222n10, 223n12
'Hippocratic' texts, corpus, medicine 3, 56, 93–5, 183, 189–95, 197–8, 200–2
Homer 140–1, 143
Hubble, Edwin 212n7
Hubble (telescope) 220n23

Julius Caesar 136, 138, 184
Juvenal 184

Kepler, Johannes 186
Koch, Robert 115
Kuhn, Thomas 67, 73

Lamarck, Jean-Baptiste; Lamarckism 40–2
Lavoisier, Antoine 80–1, 84, 86, 215n18, 217n14
Leeuwenhoek, Anton von 115
Leucippus; *see also* Democritus 12, 76, 212n9
Lister, Joseph 116
Lloyd, Geoffrey 200
Lucretius 16, 41–2, 57
Lyell, Charles 34–5, 47, 213n7

Marshall, Barry 116
Mayhew, Robert 61–2
Melissus 4
Mercator, Gerardus 142
Methodists (medical sect) 109, 179
Meton 56, 136, 138, 145, 182
miasma; *see also* pollution 113, 115
Miletus 1, 3, 141

Nature of Man, On the 91–2, 111, 189, 218n23, 222n9
Netz, Reviel 131–2
Newton, Isaac 67, 73
Newton's First Law 74
Nonius Datus 150

Oppenheimer, J. Robert 162

Pappus 153, 163
Parmenides 4, 10–11, 18, 23, 171
Pasteur, Louis 31, 115–16
Penzias, Arno 6
Persia, the Persian empire 1, 95, 219n4
Peutinger Table, the 143
Philoponus 65, 67, 70–3, 75
Plato 4, 7–8, 11, 13, 15, 20, 22, 50, 52, 55, 69, 130, 154, 157, 171, 188, 212n10, 215n1, 222n10
Pliny the Elder 95, 154, 164, 197, 221n38
Plutarch 157, 183
pollution 93–5, 97, 110–11, 113, 115
Polybius 132–3
Posidonius 147–8
Pre-Socratics (*for individual Pre-Socratics see entries by name*) 1–50, 54, 56, 59
Protagoras 16
Proust, Joseph-Louis 81–2
(Claudius) Ptolemy 3, 133, 142–3, 146–7, 184–6, 188

Ptolemy II, Ptolemaic kings 132, 156, 175, 177
Pythagoras, Pythagoreans 128–9, 164
Pythagoras' theorem 129, 132, 219n3

Rome, Roman empire 3, 95, 134, 136, 151–2, 182
  water supply of 149–50
Rose, Steven 180

Sacred Disease, On the (text) 92–4, 111, 189, 216n6, 217n20
sacred disease (condition) 94, 110–11
Sammelweis, Ignaz 116–17
Sanctorius, Sanctorio 99–100
Sceptics (New Academics) 168–9, 186
scientific method, definition of 59–63, 98
Seed, On the; Nature of the Child, On the 120–5
Sicily; Syracuse 34, 43, 131, 183, 220n26
Simplicius 72, 77
Snow, John 115–16
Socrates 1, 50, 55–6
sophists 4, 55–6
Soranus 122, 124, 107, 126–7, 175, 196
Sosigenes 136–8
Stoics; *see also* Zeno of Citium 17, 20, 24, 26, 168–9, 213, 221n43

Strabo 140, 144
Strato 76–7, 79, 83–4, 171

Thales 1–2, 32, 56, 134, 211n3
Theophrastus 58, 76, 214n23, 217n10
Thucydides 56, 109–10, 113, 115, 183
Tiberius 184
Turing, Alan 161
Turing machine 160–1

Varro 114
Vitruvius 113–14, 133, 149–50, 154
vivisection (general) 176–81, 105–7
vivisection, human 176–81
void *and* microvoid 8, 16–17, 69, 76–9, 84–6, 163

Wallace, Alfred Russell 45–9
Wilson, Robert 6

Xenophanes 34–6, 49, 214n23
Xenophon 119

Zeno of Citium; *see also* Stoics 20
Zeno of Elea 4

CPSIA information can be obtained
at www.ICGtesting.com
Printed in the USA
LVHW011345080822
725426LV00008B/272